Dynamics and Control

Stability and Control: Theory, Methods and Applications
A series of books and monographs on the theory of stability and control
Edited by A.A. Martynyuk, Institute of Mechanics, Kiev, Ukraine and V. Lakshmikantham, Florida Institute of Technology, USA

Volume 1

Theory of Integro-Differential Equations
V. Lakshmikantham and M. Rama Mohana Rao

Volume 2

Stability Analysis: Nonlinear Mechanics Equations
A.A. Martynyuk

Volume 3

Stability of Nonautonomous Systems (Method of Limiting Equations)
J. Kato, A.A. Martynyuk and A.A. Shestakov

Volume 4

Control Theory and its Applications
E.O. Roxin

Volume 5

Advances in Nonlinear Dynamics
edited by S. Sivasundaram and A.A. Martynyuk

Volume 6

Solving Differential Problems by Multistep Initial and Boundary Value Methods
L. Brugnano and D. Trigiante

Volume 7

Dynamics of Machines with Variable Mass
L. Cveticanin

Volume 8

Optimization of Linear Control Systems: Analytical Methods and Computational Algorithms
F.A. Aliev and V.B. Larin

Volume 9

Dynamics and Control
edited by G. Leitmann, F.E. Udwadia and A.V. Kryazhimskii

Dynamics and Control

Edited by

G. Leitmann
University of California
Berkeley, California, USA

F.E. Udwadia
University of Southern California
Los Angeles, California, USA

and

A.V. Kryazhimskii
Russian Academy of Sciences
Moscow, Russia

GORDON AND BREACH SCIENCE PUBLISHERS
Australia • Canada • China • France • Germany • India • Japan • Luxembourg
Malaysia • The Netherlands • Russia • Singapore • Switzerland

 Printed in Singapore.

Amsteldijk 166
1st Floor
1079 LH Amsterdam
The Netherlands

British Library Cataloguing in Publication Data

A catalogue record for this book is available from the British Library

ISBN 90-5699-172-8
ISSN 1023-6155

Contents

Dynamical Systems

Applications to Social and Environmental Problems

Introduction to the Series

The problems of modern society are both complex and interdisciplinary. Despite the apparent diversity of problems, tools developed in one context are often adaptable to an entirely different situation. For example, consider the Lyapunov's well known second method. This interesting and fruitful technique has gained increasing significance and has given a decisive impetus for modern development of the stability theory of different equations. A manifest advantage of this method is that it does not demand the knowledge of solutions and therefore has great power in application. It is now well recognized that the concept of Lyapunov-like functions and the theory of differential and integral inequalities can be utilized to investigate qualitative and quantitative properties of nonlinear dynamic systems. Lyapunov-like functions serve as vehicles to transform the given complicated dynamic systems into a relatively simpler system and therefore it is sufficient to study the properties of this simpler dynamic system. It is also being realized that the same versatile tools can be adapted to discuss entirely different nonlinear systems, and that other tools, such as the variation of parameters and the method of upper and lower solutions provide equally effective methods to deal with problems of a similar nature. Moreover, interesting new ideas have been introduced which would seem to hold great potential.

Control theory, on the other hand, is that branch of application-oriented mathematics that deals with the basic principles underlying the analysis and design of control systems. To control an object implies the influence of its behavior so as to accomplish a desired goal. In order to implement this influence, practitioners build devices that incorporate various mathematical techniques. The study of these devices and their interaction with the object being controlled is the subject of control theory. There have been, roughly speaking, two main lines of work in control theory which are complementary. One is based on the idea that a good model of the object to be controlled is available and that we wish to optimize its behavior, and the other is based on the constraints imposed by uncertainty about the model in which the object operates. The control tool in the latter is the use of feedback in order to correct for deviations from the desired behavior. Mathematically, stability theory, dynamic systems and functional analysis have had a strong influence on this approach.

Volume 1, *Theory of Integro-Differential Equations*, is a joint contribution by V. Lakshmikantham (USA) and M. Rama Mohana Rao (India).

Volume 2, *Stability Analysis: Nonlinear Mechanics Equations*, is by A.A. Martynyuk (Ukraine).

Volume 3, *Stability of Motion of Nonautonomous Systems: The Method of Limiting Equations*, is a collaborative work by J. Kato (Japan), A.A. Martynyuk (Ukraine) and A.A. Shestakov (Russia).

Volume 4, *Control Theory and its Applications*, is by E.O. Roxin (USA).

Volume 5, *Advances in Nonlinear Dynamics*, is edited by S. Sivasundaram (USA) and A.A. Martynyuk (Ukraine) and is a multiauthor volume dedicated to Professor S. Leela.

Volume 6, *Solving Differential Problems by Multistep Initial and Boundary Value Methods*, is a joint contribution by L. Brugnano (Italy) and D. Trigiante (Italy).

Volume 7, *Dynamics of Machines with Variable Mass*, is by L. Cveticanin (Yugoslavia).

Volume 8, *Optimization of Linear Control Systems: Analytical Methods and Computational Algorithms*, is a joint work by F.A. Aliev (Azerbaijan) and V.B. Larin (Ukraine).

Volume 9, *Dynamics and Control*, is edited by G. Leitmann (USA), F.E. Udwadia (USA) and A.V. Kryazhimskii (Russia).

Due to the increased interdependency and cooperation among the mathematical sciences across the traditional boundaries, and the accomplishments thus far achieved in the areas of stability and control, there is every reason to believe that many breakthroughs await us, offering existing prospects for these versatile techniques to advance further. It is in this spirit that we see the importance of the 'Stability and Control' series, and we are immensely thankful to Gordon and Breach Science Publishers for their interest and cooperation in publishing this series.

Preface

The workshops on Dynamics and Control were initiated in 1988 at the University of Southern California by Professor Janislaw M. Skowronski. Since then, they have been held annually at various locations around the globe. The underlying theme of these workshops has been the promotion of advanced methods in the understanding of dynamical systems and in control theory, with special emphasis on the development of methods dealing with systems where there are uncertainties.

The Eighth Workshop on Dynamics and Control was held in Sopron, Hungary, and was co-sponsored by the University of Southern California, the International Institute for Applied Systems Analysis, the University of California at Berkeley, the National Science Foundation and the Hungarian Academy of Sciences. This volume presents some of the papers presented at this workshop. Many of the papers constitute significant advances in our thinking about dynamical systems, their control, and the effects of uncertainties. A significant number of papers deal with expanding the scope of applicability of our knowledge of dynamics and control to social and environmental problems.

The papers included in this book can be divided into three major groups. The first group consists of papers dealing with the development of control methods. Chernousko deals with the control of dynamical systems subject to mixed constraints on both the control and state variables. Reithmeier and Leitmann study pole placement via Lyapunov theory for constrained control of systems with mismatched uncertainties. Kim and Chen develop a control design method for flexible manipulators with mismatched uncertainties. Krasovskii deals with the optimal control problem for the minimax–maximin of a given objective functional within the framework of game theory. Lukoyanov develops optimal feedback control for an uncertain system by considering a two-person zero-sum differential game with hereditary information. Gyurkovics presents a receding horizon control scheme for stabilization of discrete-time uncertain systems with bounded controllers. Mohler and Khapalov develop sufficient conditions for the complete global controllability of time-invariant nonhomogeneous bilinear systems. Mordukhovich and Zhang present results on the minimax robust control design of systems modeled by parabolic partial differential equations with uncertain perturbations under state and control constraints.

The second group of papers deals with advances in dynamical systems, and especially in game-theoretic approaches. Kurzhanski presents two alternative ways for the calculation of information sets in the state estimation problem under uncertainty. Ákos obtains an upper bound on the time constants for quadratic Lyapunov functions. Tibken, Hofer and Demir provide ways of finding regions of attraction for dynamical polynomial systems. Martynyuk obtains sufficient conditions for the stability of nonlinear equations which pertain to generalized Kolmogorov population models. Kryazhimskii and Sonnevend show that in higher dimensional bimatrix game-dynamical systems mixed Nash equilibria can occur as points of attraction. Lastly, Tarasyev gives analytic results for evolutionary non-zero sum games using control feedbacks.

The third group of papers deals with applications of dynamical systems methodology to social and environmental problems. Feichtinger considers policies for the optimal control of law enforcement. Svirezhev and von Bloch introduce the concept of virtual biospheres in modeling climate change in terms of dynamical systems. Raju and Udwadia consider the global dynamics of coupled exponential maps commonly found in ecological modeling. Petrosjan deals with developing strategies for electing directorial councils in a game-theoretic framework. And finally, Kleimenov presents a new approach to building a 2×2 repeated bimatrix game wherein players with various behavior types are considered.

The organizing committee of the Eighth Workshop thanks all the authors for their active cooperation and participation during the preparation of this book. The organizers are appreciative of the financial support provided for this workshop by the International Institute for Applied System Analysis, the National Science Foundation and the Hungarian Academy of Sciences.

Firdaus E. Udwadia
Arkadii V. Kryazhimskii
George Leitmann

List of Contributors

László Ákos
Department of Operations Research
Eötvös University
Muzeum k. 6–8
H-1088 Budapest
Hungary

Y.-H. Chen
The George W. Woodruff School of Mechanical Engineering
Georgia Institute of Technology
Atlanta, GA 30332-0405
USA

Felix L. Chernousko
Department of Control
Institute for Problems in Mechanics
Russian Academy of Sciences
pr. Vernadskogo 101
Moscow 117526
Russia

C. Demier
Department of Measurement, Control and Microtechnology
University of Ulm
D-8906 Ulm
Germany

Gustav Feichtinger
Vienna University of Technology
Institute for Econometrics, Operations Research and Systems Theory
Vienna
Austria

Éva Gyurkovics
Department of Mathematics

Technical University of Budapest
Faculty of Mechanical Engineering
Műegyetem rkp. 3
H-1521 Budapest
Hungary

E.P. Hofer
Department of Measurement, Control and Microtechnology
Faculty of Engineering Sciences
University of Ulm
PO Box 4066
D-7900 Ulm
Germany

D.H. Kim
The George W. Woodruff School of Mechanical Engineering
Georgia Institute of Technology
Atlanta, GA 30332-0405
USA

Alex Y. Khapalov
Department of Electrical and Computer Engineering
Oregon State University
Cornvallis, OR 97331-3211
USA

Anatolii F. Kleimenov
Institute of Mathematics and Mechanics
Ural Branch of Russian Academy of Sciences
S. Kovalevskoi 16
Ekaterinburg 620219
Russia

Andrew N. Krasovskii
Department of Mathematics and Mechanics
Urals State University
Lenina 51
Ekaterinburg 620083
Russia

Arkadii V. Kryazhimskii
Mathematical Steklov Institute

Russian Academy of Sciences
Gubkina 8, CSPI
Moscow 117966
Russia

A.B. Kurzhanski
Faculty of Computational Mathematics and Cybernetics (VMK)
Moscow State University (MGU)
Moscow 119899
Russia

George Leitmann
College of Engineering
University of California, Berkeley
232 Hearst Memorial Mining Building
Berkeley, CA 94720
USA

Nikolai Yu. Lukoyanov
Department of Dynamic Systems
Institute of Mathematics and Mechanics
Ural Branch of Russian Academy of Sciences
S. Kovalevskoi 16
Ekaterinburg 620219
Russia

Anatoli A. Martynyuk
Department of Stability of Processes
Institute of Mechanics
National Ukrainian Academy of Sciences
Nesterov str. 3
152057 Kiev 57
Ukraine

R.R. Mohler
Department of Electrical and Computer Engineering
Oregon State University
Cornvallis, OR 97331
USA

Boris S. Mordukhovich
Department of Mathematics

Wayne State University
Detroit, MI 48202
USA

Leon A. Petrosjan
Robotechnic Laboratory
Faculty of Applied Mathematics
St Petersburg State University
Bibliotechnaya pl. 2, Petrodvoretz
St Petersburg 198904
Russia

Narayanan Raju
Department of Aerospace Engineering
University of Southern California
854 W. 36th Pl.
Los Angeles, CA 90007
USA

Eduard Reithmeier
Institut für Mess-und Regelungstechnik
Universität Hannover
Hannover
Germany

†György Sonnevend
Department of Numerical Analysis
Eötvös University
Muzeum k. 6-8
H-1088 Budapest
Hungary

Yuri Svirezhev
Department of Integrated Systems Analysis
Potsdam-Institute for Climate Impact Research
PO Box 601203
D-14412 Potsdam
Germany

A.M. Tarasyev
Dynamic Systems Department
Institute of Mathematics and Mechanics

Ural Branch of Russian Academy of Sciences
S. Kovalevskoi 16
Ekaterinburg 620219
Russia

B. Tibken
Department of Measurement, Control and Microtechnology
University of Ulm
D-8906 Ulm
Germany

Firdaus E. Udwadia
Department of Mechanical Engineering
University of Southern California
430K Olin Hall
Los Angeles, CA 90089-1453
USA

Werner von Bloh
Department of Integrated Systems Analysis
Potsdam Institute for Climate Impact Research
PO Box 601203
D-14412 Potsdam
Germany

Kaixia Zhang
Department of Mathematics
Wayne State University
Detroit, MI 48202
USA

CONTROL METHODOLOGY

1 CONTROL OF DYNAMICAL SYSTEMS SUBJECT TO MIXED CONSTRAINTS

FELIX L. CHERNOUSKO

Institute for Problems in Mechanics, Russian Academy of Sciences, Moscow, Russia

1 Introduction

Mixed constraints imposed on both state and control variables of a dynamical system, as well as integral constraints imposed on these variables, are often present in applications. For instance, if a system includes an electric drive, we are to take into account constraints imposed on the angular velocity of the rotor, on the control torques created by the motor, and also on the combination of these variables. Energy and heat restrictions are usually reduced to integral constraints. All these constraints are essential for robots controlled by electric actuators. Different mixed, state, and integral constraints arise also in various control problems for ecological, economical, and mechanical systems. It is well known that state and mixed constraints present considerable difficulties in the frames of the optimal control theory. Solutions of optimal control problems under such constraints can be usually obtained only numerically for given sets of initial and boundary conditions. In this paper, we extend the well-known Kalman's method [1, 2] originally developed for linear systems without control constraints to systems subject to mixed, state, and integral constraints. In our earlier papers [3–5], we took into account only control constraints. In Kalman's approach, the open-loop control is formed as a linear combination of the natural modes of the system. We derive sufficient controllability conditions which ensure that the obtained control satisfies all imposed constraints and brings our system to the prescribed terminal state in finite time. The terminal time is not fixed a priori and is to be chosen so that all constraints are fulfilled. Thus, our approach is semi-analytical: the control structure is obtained explicitly, whereas some parameters are to be calculated numerically. The proposed technique is applied to a dynamical system of the fourth order which is a model for mechanical systems controlled by electric drives.

2 Statement of the Problem

Consider a linear controlled system described by the equation

$$\dot{x} = A(t)x + B(t)u + f(t). \tag{1}$$

Here $x = (x_1, \ldots, x_n)$ is the n-vector of state, $u = (u_1, \ldots, u_n)$ is the m-vector of control. The matrices A and B as well as the n-vector f are given piecewise continuous functions of time t. Let the following constraints be imposed on x and u:

$$\begin{aligned} &|C^i(t)x(t) + D^i(t)u(t) + \int_{t_0}^{T} \left[G^i(t,\tau)x(\tau) + H^i(t,\tau)u(\tau)\right] d\tau \\ &\quad + \mu^i(t)| \le 1, \quad i = 1, \ldots, r, \end{aligned} \tag{2}$$

$$\begin{aligned} &\left(p^j(t), x(t)\right) + \left(q^j(t), u(t)\right) + \int_{t_0}^{T} \left\{\left(g^j(t,\tau), x(\tau)\right)\right. \\ &\left. + \left(h^j(t,\tau), u(\tau)\right)\right\} d\tau \le 1, \quad j = 1, \ldots, s. \end{aligned}$$

Here and below, scalar products of vectors are denoted by (,). The constraints (2) are to be fulfilled for all $t \in [t_0, T]$ where t_0 and T are the initial and terminal time instants, respectively. The instant t_0 is fixed, whereas T is not fixed. The $l \times n$-matrices C^i and G^i, $l \times m$-matrices D^i and H^i, l-vector μ^i, n-vectors p^j and g^j and m-vectors q^j and h^j are given piecewise continuous functions of $t \in [t_0, T]$ and $\tau \in [t_0, T]$. The constraints (2) imply restrictions on the absolute values and components of certain linear combinations of x, u and some integrals of these variables. These constraints include the most commonly encountered restrictions imposed on control, state, and their combinations. Let the initial and terminal states be fixed

$$x(t_0) = x^0, \quad x(T) = x^1. \tag{3}$$

Denote by $\Phi(t)$ the fundamental matrix of (1) defined by

$$\dot{\Phi} = A\Phi, \quad \Phi(t_0) = E \tag{4}$$

where E is the unit $n \times n$-matrix. The solution of our system (1) satisfying the initial condition (3) is given by

$$x(t) = \Phi(t)\left\{x^0 + \int_{t_0}^{t} \Phi^{-1}(\tau)[B(\tau)u(\tau) + f(\tau)]d\tau\right\}. \tag{5}$$

Substituting (5) into the terminal condition (3), we obtain

$$\int_{t_0}^{T} \Phi^{-1}(t)B(t)u(t)dt = x^* \tag{6}$$

$$x^* = \Phi^{-1}(T)x^1 - x^0 - \int_{t_0}^{T} \Phi^{-1}(t)f(t)dt \tag{7}$$

We are to find a control $u(t)$ satisfying (2) and (6).

3 Design of Control

According to [1], we put

$$u(t) = Q^T(t)\,C, \quad Q(t) = \Phi^{-1}(t)\,B \tag{8}$$

where C is a constant n-vector, Q is an $n \times m$-matrix, and ^{T} denotes the transposed matrix. Substituting (8) into (6), we obtain the equation for C

$$R(T)C = x^*, \quad R(T) = \int_{t_0}^{T} Q(\tau)\,Q^T(\tau)d\tau \tag{9}$$

where $R(t)$ is a symmetric non-negative definite $n \times n$-matrix defined for $t \leq t_0$. We assume that $R(t)$ is positive definite, i.e. that the system (1) is controllable [1, 2]. Then the equation (9) for C has a unique solution

$$C = R^{-1}(T)\,x^*. \tag{10}$$

Using (7)–(10), we present the control and state (5) as follows

$$\begin{gathered} x(t) = \Phi(t)\left[\Phi^{-1}(T)x^1 + R_1(t,T)x^* - \int_{t}^{T} \Phi^{-1}(\tau)f(\tau)d\tau\right] \\ R_1(t,T) = R(t)R^{-1}(T) - E, \; u(t) = Q^T(t)R^{-1}(T)x^* \end{gathered} \tag{11}$$

Substituting (11) into the constraints (2), we rewrite them in the form

$$\begin{gathered} |F^i(t,T)x^* + \varphi^i(t,T)| \leq 1, \quad i = 1,\ldots,r, \\ (\psi^j(t,T), x^*) + \chi^j(t,T) \leq 1, \quad j = 1,\ldots,s. \end{gathered} \tag{12}$$

Here, the matrices and vectors F^i, φ^i, ψ^j, and χ^i are easily expressed through the given matrices and vectors from (2), and the defined above matrices Φ, Q, R, and R_1. The inequalities (12) which should hold for all $t \in [t_0, T]$ impose constraints on x^* and T. These constraints can be regarded as controllability conditions. Let us simplify (12) assuming that

$$\begin{gathered} |\varphi^i(t,T)| \leq \varphi_0^i < 1, \quad i = 1,\ldots,r, \\ \chi^j(t,T) \leq \chi_0^j < 1, \quad j = 1,\ldots,s \end{gathered} \tag{13}$$

where φ_0^i and χ_0^j are constants. Using (13) and Cauchy's inequality, we replace (12) by the conditions

$$|x^*| \leq \min_i \left\{ (1-\varphi_0^i) \left[\max_t \sum_{j=1}^{l} \sum_{k=1}^{n} \left(F_{jk}^i(t,T) \right)^2 \right]^{-1/2} \right\}$$
$$|x^*| \leq \min_j \left\{ (1-\chi_0^i) \left[\max_t |\psi^j(t,T)| \right]^{-1} \right\} \qquad (14)$$
$$i=1,\dots,r, \quad j=1,\dots,s, \quad t \in [t_0,T].$$

According to (7), the sufficient controllability conditions (14) impose constraints on x^0, x^1, and T. If (12) or (14) hold for some T, then the control u defined by (11) satisfies all constraints and brings our system (1) from the state x^0 at $t=t_0$ to the terminal state x^1 at $t=T$. Of course, the conditions (12) can be simplified also in other ways different from (14); one such possibility is used below.

4 Model of an Electromechanical System

Consider a system with two degrees of freedom described by equations

$$m_1\ddot{\xi}_1 = c(\xi_2-\xi_1)+F, \quad m_2\ddot{\xi}_2 = c(\xi_1-\xi_2). \qquad (15)$$

Here, ξ_1 and ξ_2 are the coordinates of the masses m_1 and m_2, respectively, c is the stiffness of the spring connecting these masses, F is the control force (or torque) created by an electric motor and applied to the first mass. Some mechanical systems described by equations (15) are shown in Fig. 1.

We have $F=k_1I$, $RI+k_2\dot{\xi}_1=U$, where I is the electric current in the rotor circuit of the motor, U is the voltage, R is the electric resistance, and k_1 and k_2 are constants. The constraints imposed on the angular velocity of the motor, torque, and voltage can be expressed as follows

$$|\dot{\xi}_1| \leq U_0k_2^{-1}, \quad |F| \leq F_0, \quad |U| \leq U_0 \qquad (16)$$

where U_0 and F_0 are constants. Equations (15) and constraints (16) describe various mechanical systems with two degrees of freedom controlled by electric drives. Introducing dimensionless and normalized variables and constants

$$t'=\omega t, \quad x_1=\frac{m_1\xi_1+m_2\xi_2}{(m_1+m_2)l_0}, \quad x_2=\frac{m_1\dot{\xi}_1+m_2\dot{\xi}_2}{(m_1+m_2)l_0\omega},$$
$$x_3=\frac{m_1\xi_1-m_2\xi_2}{(m_1+m_2)l_0}, \quad x_4=\frac{m_1\dot{\xi}_1-m_2\dot{\xi}_2}{(m_1+m_2)l_0\omega}, \qquad (17)$$
$$u=\frac{F}{(m_1+m_2)l_0\omega^2}, \quad \omega^2=\frac{c(m_1+m_2)}{m_1m_2}, \quad l_0=\frac{F_0m_1m_2}{c(m_1+m_2)^2},$$
$$p=l_0\omega k_2U_0^{-1}, \quad q=(m_1+m_2)l_0\omega^2Rk_1^{-1}U_0^{-1}, \quad \mu=m_2m_1^{-1},$$

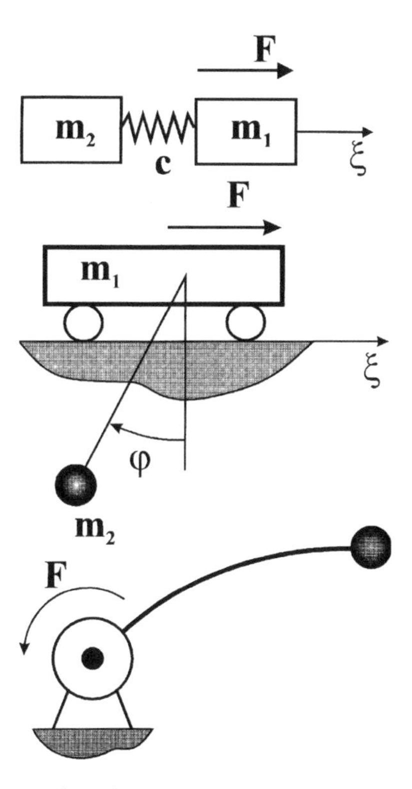

Figure 1 Examples of mechanical systems described by equations (15).

we reduce equations (15) and constraints (16) to the form

$$\dot{x}_1 = x_2, \quad \dot{x}_2 = u, \quad \dot{x}_3 = x_4, \quad \dot{x}_4 = -x_3 + u, \tag{18}$$

$$|u| \leq 1, \quad p|x_2 + \mu x_4| \leq 1, \quad |p(x_2 + \mu x_4) + qu| \leq 1. \tag{19}$$

Here, dots denote derivatives with respect to the dimensionless time t' which is denoted below by t. The control $u(t)$ should satisfy the mixed constraints (19) and bring the system (18) from the initial state

$$x_1(0) = x_1^0, \quad x_2(0) = x_2^0, \quad x_3(0) = x_3^0, \quad x_4(0) = x_4^0 \tag{20}$$

to the zero terminal state $x_i(T) = 0$, $i = 1, 2, 3, 4$, where T is not fixed.

5 Control for the Model

A control for our electromechanical model can be designed as proposed in Section 3. Omitting cumbersome calculations and estimates simplifying the inequalities (12), we present only final results. For simplicity, we take $T = 2\pi k$ where the integer k is to be chosen. The control (11) is given by

$$\begin{aligned} u(t) &= (W(t,T), x^0)\Delta^{-1}, \quad \Delta = T(T^2 - 24), \\ W_1 &= 12t - 6T + 24\sin t, \quad W_2 = 6tT - 4T^2 + 24 + 12T\sin t, \\ W_3 &= 24t - 12T + 2T^2 \sin t, \quad W_4 = -2(T^2 - 24)\cos t \end{aligned} \tag{21}$$

where W_i are the components of the 4-vector W and x^0 is the initial 4-vector (see (20)). Finally, we reduce the constraints (19) to the form different from (14) and containing the absolute values $|x_i^0|$ from (20). We obtain

$$\begin{aligned} &\sum_{i=1}^{4} A_i |x_i^0| \le 1, \\ &p\left\{D_1|x_1^0| + D_2|x_2^0| + D_3\left[(x_3^0)^2 + (x_4^0)^2\right]^{1/2}\right\} + q\sum_{i=1}^{4} A_i|x_i^0| \le 1 \end{aligned} \tag{22}$$

where $A_i(T)$ and $D_i(T)$ are given by

$$\begin{aligned} A_1 &= 6(T+4)\Delta^{-1}, \quad A_2 = 4(T^2 + 3T - 6)\Delta^{-1}, \\ A_3 &= 2T(T+6)\Delta^{-1}, \quad A_4 = 2T^{-1}, \\ D_1 &= \{\max[3T^2/2, 24(2+\mu)] + 6\mu T\}\Delta^{-1}, \\ D_2 &= \left[T^3 + 4\mu T^2 + 12(2+\mu)T + 24\mu\right]\Delta^{-1}, \\ D_3 &= \mu + \left[(4T^2 + 12\mu T + 48\mu)^2 + (2+\mu)^2(T^2 - 24)^2\right]^{1/2}\Delta^{-1}. \end{aligned} \tag{23}$$

Here, p, q, and μ are defined in (17). The inequalities (22) imposed on the initial state and the terminal time $T = 2\pi k$ are sufficient controllability conditions for our control problem (18)–(20). Note that all A_i, $i = 1, 2, 3, 4$, and D_1 tend to zero whereas $D_2 \to 1$, $D_3 \to \mu$ as $T \to \infty$. Therefore, both conditions (22) are fulfilled for sufficiently large T, if

$$p\left\{|x_2^0| + \mu\left[(x_3^0)^2 + (x_4^0)^2\right]^{1/2}\right\} < 1 \tag{24}$$

If (24) holds, then there exists $T = 2\pi k$ such that the control (21) solves our problem. To choose T, we take $k = 1, 2, \ldots$, calculate A_i and D_i from (23), and verify (22). The smallest $T = 2\pi k$ for which both inequalities (22) hold can be taken as the terminal time. Thus, our control problem is solved by simple and explicit calculations. Computer simulation shows quite satisfactory behavior of the system under the obtained control.

Though the control (21) is an open-loop one, it can also be used for a feedback control. To do that, we should, in certain time intervals, regard x^0 in (21) as a current state and recalculate the terminal time T as described above.

6 Numerical Example

Let us consider a numerical example. The control (21) was applied to the system (18) subject to the constraints (19). The dimensionless parameters (17) in the constraints (19) were taken as follows

$$p = 0.1, \quad q = 0.5, \quad \mu = 0.5. \tag{25}$$

We choose the following initial conditions (20)

$$x_1^0 = -5, \quad x_2^0 = 0, \quad x_3^0 = 5, \quad x_4^0 = 0. \tag{26}$$

It can be easily verified that the parameters (25) and the initial data (26) satisfy the sufficient controllability condition (24). The numerical realization of the procedure described in Section 5 provides the minimal integer k under which both the conditions (22) are satisfied. We have

$$k = 3, \quad T = 6\pi. \tag{27}$$

After that, we obtain numerically the control $u(t)$ according to (21) and the state trajectory $x(t)$. Some results are depicted in Figs. 2, 3.

Fig. 2 presents the time histories of the functions

$$u, \quad p(x_2 + \mu x_4), \quad p(x_2 + \mu x_4) + qu \tag{28}$$

shown by the curves 1, 2, and 3, respectively. One can see that all the imposed constraints (19) are satisfied with a considerable margin.

The projection of the four-dimensional state trajectory $x(t)$ onto the planes (x_1, x_2) and (x_3, x_4) is shown in Fig. 3 by the curves 1 and 2, respectively. Both curves reach the point (0, 0) at the same time $t = T = 6\pi$, see (27).

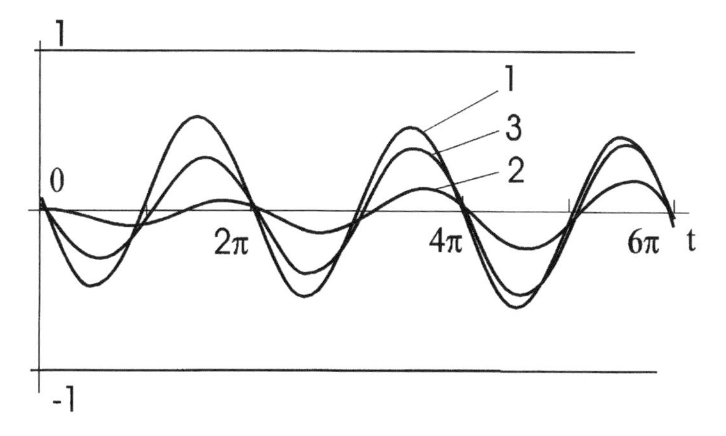

Figure 2 Time histories of the functions (28) in the constraints (19).

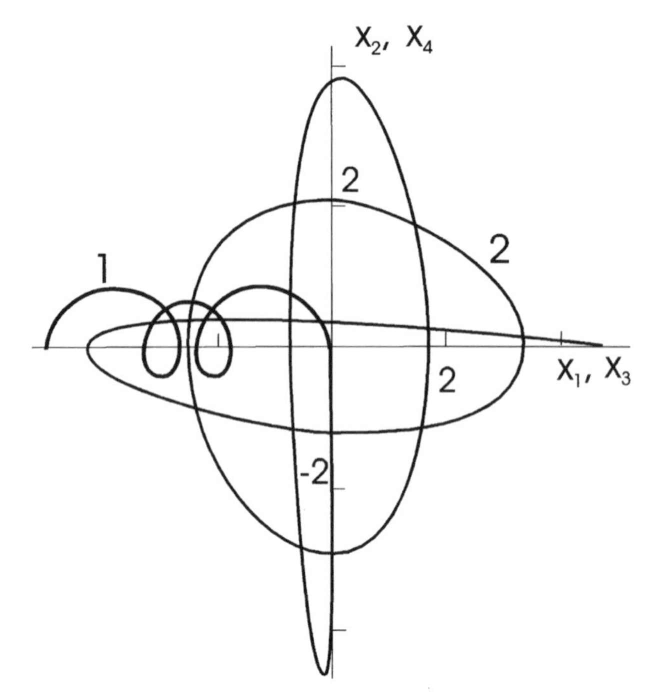

Figure 3 Projections of the state trajectory $x(t)$ onto the planes (x_1, x_2) and (x_3, x_4) (curves 1 and 2, respectively).

7 Conclusion

Kalman's approach is extended to the case of linear controlled systems subject to various control, state, mixed, and integral constraints. The control is presented in an explicit analytical form, and the terminal time can be obtained by a simple procedure.

References

[1] Kalman, R.E. (1960) On the general theory of control systems, in *Proc. 1st IFAC Congress*, Butterworth, London, 481–500.
[2] Kalman, R.E. *et al.* (1968) *Topics in Mathematical System Theory*. McGraw-Hill.
[3] Chernousko, F.L. (1988) On the construction of a bounded control in oscillatory systems. *J. Appl. Math. and Mech.* (PMM), **52**(4), 426–433.
[4] Chernousko, F.L. (1990) Constrained controls in linear oscillating systems, in *Analysis and Optimization of Systems, Proc. 9th International Conference*, Lecture Notes in Control and Information Sciences, N 144 (A. Bensoussan and J.L. Lions, Eds.). Springer-Verlag, Berlin, 580–589.
[5] Dobrynina, I.S. and Chernousko, F.L. (1994) Constrained control of a fourth-order linear system. *J. Computer and Systems Science International*, **32**(4), 108–115.

2 POLE PLACEMENT VIA LYAPUNOV FOR CONSTRAINED CONTROL OF MISMATCHED SYSTEMS

EDUARD REITHMEIER[1] and GEORGE LEITMANN[2]

[1]*Institut für Mess- und Regelungstechnik, Universität Hannover, Hannover, Germany*
[2]*College of Engineering, University of California at Berkeley, Berkeley, CA 94720, USA*

1 Introduction

The main objective discussed in [5] is the suppression of undesired noise or vibrations in dynamical systems which are modelled by an o.d.e. of the form

$$\dot{\boldsymbol{x}} = \boldsymbol{A}\boldsymbol{x} + \boldsymbol{B}(\boldsymbol{x})\boldsymbol{u} + \boldsymbol{e}(\boldsymbol{x}, t) \quad \text{with} \quad \boldsymbol{x}(0) = \boldsymbol{x}_0 \in \mathbb{R}^n. \tag{1}$$

The variable $\boldsymbol{x} \in \mathbb{R}^n$ describes the state of the system. $\boldsymbol{A} \in \mathbb{R}^{n,n}$ is assumed to be a constant and stable system matrix. $\boldsymbol{u} \in \mathbb{R}^m$ with $u_k \in [-1, +1]$ for $k = 1, \ldots, m$ is the constrained control input. The input matrix $\boldsymbol{B} \in C^0[\mathbb{R}^n, \mathbb{R}^m]$ may be state-dependent. The *Caratheodory function* $\boldsymbol{e} : \mathbb{R}^n \times \mathbb{R} \to \mathbb{R}^n$ models time-dependence, additional nonlinearities, and unknown disturbances. However, it is assumed to be uniformly bounded; that is, $\|\boldsymbol{e}(\boldsymbol{x}, t)\| \leq \eta$ for all $(\boldsymbol{x}, t) \in \mathbb{R}^n \times \mathbb{R}$, where the constant $\eta \in (0, \infty)$ may be unknown.

In [5, 6] the authors showed that the control design method based on Lyapunov stability theory leads to a unique control function

$$u_k = p_k(\boldsymbol{x}) = -\operatorname{sgn}[b_k(\boldsymbol{x})], \quad k \in \{1, \ldots, m\} \tag{2}$$

that minimizes the time derivative $\dot{V}(\boldsymbol{x}(t))$ of any given Lyapunov function candidate $V \in C^1[\mathbb{R}^n, \mathbb{R}]$ along any trajectory $t \mapsto \boldsymbol{x}(t)$ which satisfies equation (1) with control (2). The indicator function b_k in that case is given by

$$b_k(\boldsymbol{x}) = \sum_{j=1}^{n} \left[\frac{\partial V}{\partial x_j}(\boldsymbol{x}) \cdot B_{jk}(\boldsymbol{x}) \right] \quad \text{where} \quad B_{jk}(\boldsymbol{x}) := \boldsymbol{e}_j^T \boldsymbol{B}(\boldsymbol{x}) \tilde{\boldsymbol{e}}_k. \tag{3}$$

$$(\boldsymbol{e}_i^T \boldsymbol{e}_j = \delta_{ij},\ \tilde{\boldsymbol{e}}_i^T \tilde{\boldsymbol{e}}_j = \delta_{ij},\ \boldsymbol{e}_i \in \mathbb{R}^n,\ \tilde{\boldsymbol{e}}_i \in \mathbb{R}^m).$$

As shown in [6], the commonly used Lyapunov function candidate $V(\boldsymbol{x}) = \boldsymbol{x}^T \boldsymbol{P} \boldsymbol{x}$ with $\boldsymbol{P} \in \mathbb{R}^{n,n}$ and $\boldsymbol{P} > 0$ is actually a Lyapunov function if we determine $\boldsymbol{P}$ via the algebraic Lyapunov equation $PA + A^T P = -Q$ where $Q \in \mathrm{IR}^{n,n}$ is any given positive definite matrix. In that case, the radius ρ of the ball of ultimate boundedness is given by

$$\rho = 2 \cdot \eta \cdot \frac{\lambda_{max}(P)}{\lambda_{min}(Q)} \cdot \sqrt{\frac{\lambda_{max}(P)}{\lambda_{min}(Q)}}. \tag{4}$$

$\lambda_{max}(P)$ and $\lambda_{min}(P)$ denote the maximum and minimum eigenvalues of $P > 0$. The same holds for $Q > 0$. At this point it should be noted that minimization of the Lyapunov derivative

$$\mathcal{L}_{(x,t)}[u] := x^T P(Ax + Bu + e(x, t)] \tag{5}$$

has also a geometric interpretation in IR^n: Since $P > 0$ there is an invertible $T \in \mathrm{IR}^{n,n}$ such that $P = T^T T$ or, along a trajectory of (1),

$$\begin{aligned} \mathcal{L}_{(\mathbf{x},t)}[\mathbf{u}] &= x^T T^T T\dot{x} \\ &= (Tx)^T \frac{d}{dt}(Tx), \quad \left[\frac{d}{dt}(Tx) = T\dot{x}\right] \\ &= y^T \dot{y}, \quad [y := Tx]. \end{aligned} \tag{6}$$

Thus, the proposed control (2) minimizes the inner product between $y(t)$ and $\dot{y}(t)$; in other words, $\dot{y}(t)$ points as closely as possible towards the origin. Of course, this tendency and hence the system performance depends on the choice of T or P, respectively. In Section 3, in order to illustrate further the "optimality" of the proposed control (2), we employ a T such that

$$TAT^{-1} = \Omega, \tag{7}$$

where the real matrix Ω is defined by

$$\Omega := \left[\begin{array}{ccc|ccc} \begin{pmatrix} -\delta_1 & -\omega_1 \\ \omega_1 & -\delta_1 \end{pmatrix} & & 0 & & & \\ & \ddots & & & 0 & \\ 0 & & \begin{pmatrix} -\delta_{n_c} & -\omega_{n_c} \\ \omega_{n_c} & -\delta_{n_c} \end{pmatrix} & & & \\ \hline & & & -\mu_1 & & 0 \\ & 0 & & & \ddots & \\ & & & 0 & & -\mu_{n_r} \end{array}\right] \tag{8}$$

and

- $\lambda_{2k-1} = -\delta_k + i\omega_k$ and $\lambda_{2k} = -\delta_k - i\omega_k$ are the complex eigenvalues of A
- $\mu_1, \ldots, \mu_{n_r}$ are the real eigenvalues of A.

Furthermore, because of

$$\begin{aligned}
-\boldsymbol{Q} &= \boldsymbol{PA} + \boldsymbol{A}^T\boldsymbol{P} \\
&= \boldsymbol{T}^T\boldsymbol{TA} + \boldsymbol{A}^T\boldsymbol{T}^T\boldsymbol{T} \\
&= \boldsymbol{T}^T(\boldsymbol{TAT}^{-1})\boldsymbol{T} + \boldsymbol{T}^T(\boldsymbol{T}^T)^{-1}\boldsymbol{A}^T\boldsymbol{T}^T\boldsymbol{T} \\
&= \boldsymbol{T}^T\boldsymbol{\Omega T} + \boldsymbol{T}^T(\boldsymbol{TAT}^{-1})^T\boldsymbol{T}, && (\boldsymbol{\Omega} = \boldsymbol{TAT}^{-1}) \\
&= \boldsymbol{T}^T(\boldsymbol{\Omega} + \boldsymbol{\Omega}^T)\boldsymbol{T} \\
&= -\boldsymbol{T}^T\Delta\boldsymbol{T}, && (\Delta := -(\boldsymbol{\Omega} + \boldsymbol{\Omega}^T)) \\
&= -\left(\sqrt{\Delta}\boldsymbol{T}\right)^T\left(\sqrt{\Delta}\boldsymbol{T}\right) \\
&= -\boldsymbol{R}^T\boldsymbol{R}, && \left(\boldsymbol{R} := \sqrt{\Delta}\boldsymbol{T}\right)
\end{aligned} \tag{9}$$

or

$$\boldsymbol{Q} = \boldsymbol{R}^T\boldsymbol{R} \tag{10}$$

respectively, this $\boldsymbol{T}$ defines a Lyapunov function, and (cf. [6]) the radius ρ of the ball of ultimate boundedness is given by

$$\rho = \frac{\eta}{\delta_{max}} \cdot \sqrt{\frac{\lambda_{max}(\boldsymbol{T}^T\boldsymbol{T})}{\lambda_{min}(\boldsymbol{T}^T\boldsymbol{T})}} \tag{11}$$

with

$$\delta_{max} := \max\{\delta_1, \ldots, \delta_{n_c}, \mu_1, \ldots, \mu_{n_r}\} \tag{12}$$

2 Lyapunov Approach as Limit Case of a Linear Constrained Control

The objective of this section is to show that for $\boldsymbol{B}(\boldsymbol{x}) = \boldsymbol{B} = const.$ the controller design discussed in Section 1 and in [5, 6] is a limit case of pole placement within the set of all *admissible* linear constrained feedback controllers. The attribute *admissible* denotes the restriction of $\boldsymbol{u}$ to the set of constrained linear controllers

$$\mathcal{U}_a := \{\boldsymbol{u} \in C^0[\mathbb{R}^n, \mathbb{R}^m] \mid u_k(\boldsymbol{x}) = -\mathrm{sat}(\tilde{\boldsymbol{e}}_k^T\boldsymbol{Kx});\ \ \boldsymbol{K} \in \mathbb{R}^{m,n};\ \ k \in \{1, \ldots, m\}\} \tag{13}$$

with

$$\mathrm{sat}(f(\boldsymbol{x})) := \begin{cases} +1 & \text{if } f(\boldsymbol{x}) < -1 \\ f(\boldsymbol{x}) & \text{if } -1 \le f(\boldsymbol{x}) \le 1; \quad f \in C^0[\mathbb{R}^n, \mathbb{R}] \\ -1 & \text{if } f(\boldsymbol{x}) > 1 \end{cases} \tag{14}$$

Of course, the controller $\boldsymbol{p}(\boldsymbol{x})$ described in Section 1 does not belong to $\mathcal{U}_a$. However, there exists a continuous parameter deformation

$$\begin{aligned} \tilde{p}_k(\boldsymbol{x},\cdot): \quad]0,\frac{\pi}{2}[&\to \mathcal{U}_a \\ \alpha &\mapsto -\operatorname{sat}[\tan(\alpha)\tilde{\boldsymbol{e}}_k^T\boldsymbol{B}^T\boldsymbol{P}\boldsymbol{x}] \end{aligned} \tag{15}$$

($k \in \{1, \ldots, m\}$) such that

$$p_k(\boldsymbol{x}) = \lim_{\alpha\to\pi/2} \tilde{p}_k(\boldsymbol{x},\alpha) \tag{16}$$

Suppose that a solution $t \mapsto \boldsymbol{x}(t)$ of (1) with $u_k = \tilde{p}_k(\boldsymbol{x},\alpha)$ is such that there is an interval $[t_1, t_2]$ during which not all components of $\boldsymbol{u}$ are saturated. Let

$$\boldsymbol{u} = \begin{bmatrix} \boldsymbol{u}_I \\ \boldsymbol{u}_{II} \end{bmatrix}, \quad \boldsymbol{B} = [\boldsymbol{B}_I, \boldsymbol{B}_{II}] \Longrightarrow \boldsymbol{B}\boldsymbol{u} = \boldsymbol{B}_I\boldsymbol{u}_I + \boldsymbol{B}_{II}\boldsymbol{u}_{II} \tag{17}$$

where $\boldsymbol{u}_I$ denotes the unsaturated part of $\boldsymbol{u}$. Then the systems behavior on $[t_1, t_2]$ is governed by

$$\begin{aligned} \dot{\boldsymbol{x}} &= \boldsymbol{A}\boldsymbol{x} + \boldsymbol{B}_I\boldsymbol{u}_I + \boldsymbol{B}_{II}\boldsymbol{u}_{II} + \boldsymbol{e}(\boldsymbol{x},t) \\ &= (\boldsymbol{A} - \tan(\alpha)\boldsymbol{B}_I\boldsymbol{B}_I^T\boldsymbol{P})\boldsymbol{x} + \boldsymbol{B}_{II}\boldsymbol{u}_{II} + \boldsymbol{e}(\boldsymbol{x},t) \end{aligned} \tag{18}$$

That is, as in the Lyapunov approach of [5, 6], the controller design does not take care of the uncertain excitation $\boldsymbol{e}$, but rather improves the behavior of the *nominal* system and hence its behavior in the presence of disturbances.

With respect to (17), any arbitrary but fixed $\alpha \in]0, \frac{\pi}{2}[$ determines a pole distribution of the *nominal* system. These poles are the eigenvalues of

$$\tilde{\boldsymbol{A}}(\alpha) := \boldsymbol{A} - \tan(\alpha)\boldsymbol{B}_I\boldsymbol{B}_I^T\boldsymbol{P}. \tag{19}$$

In order to investigate the damping behavior of the controlled *nominal* system, we will take a look at one of the invariants of $\tilde{\boldsymbol{A}}$ which gives information about the real parts of the eigenvalues:

$$\begin{aligned} \tilde{\boldsymbol{A}}(\alpha) &= \operatorname{tr}(\boldsymbol{A} - \tan(\alpha)\boldsymbol{B}_I\boldsymbol{B}_I^T\boldsymbol{P}) \\ &= \operatorname{tr}(\boldsymbol{A}) - \tan(\alpha)\operatorname{tr}(\boldsymbol{B}_I\boldsymbol{B}_I^T\boldsymbol{T}^T\boldsymbol{T}) \\ &= \operatorname{tr}(\boldsymbol{A}) - \tan(\alpha)\operatorname{tr}(\boldsymbol{T}\boldsymbol{B}_I\boldsymbol{B}\boldsymbol{I}^T\boldsymbol{T}^T) \\ &= -\sum_{k=1}^{n_c} 2\delta_k - \sum_{k=1}^{n_r} \mu_k - \tan(\alpha)\sum_{k=1}^{m_I} \sigma_k^2 \end{aligned} \tag{20}$$

where

$$\sigma_k^2 := \boldsymbol{e}_k^T\boldsymbol{T}\boldsymbol{B}_I\boldsymbol{B}_I^T\boldsymbol{T}^T\boldsymbol{e}_k^T. \tag{21}$$

That is, if $\alpha \to \frac{\pi}{2}$ then

$$\begin{aligned} &1) \quad \tilde{\boldsymbol{p}}(\boldsymbol{x}, \alpha) \to \boldsymbol{p}(\boldsymbol{x}) \\ &2) \quad \sum_{k=1}^{n} \mathrm{Re}\{\lambda_k(\boldsymbol{A} - \tan(\alpha)\boldsymbol{B}_I\boldsymbol{B}_I^T\boldsymbol{P})\} \to -\infty \end{aligned} \tag{22}$$

In other words, the proposed Lyapunov approach leads to the strongest possible damping "on the average".

3 Test Example and Numerical Results

To illustrate the result in Section 2, we will employ a test example already used in [6], given by the system matrix $\boldsymbol{A}$

$$\boldsymbol{A} = \begin{bmatrix} 0 & 0 & 1 & 0 \\ 0 & 0 & 0 & 1 \\ -\frac{1}{m_1}[k_1 + k_2] & \frac{k_2}{m_1} & -\frac{1}{m_1}[c_1 + c_2] & \frac{c_2}{m_1} \\ \frac{k_2}{m_2} & -\frac{k_2}{m_2} & \frac{c_2}{m_2} & -\frac{c_2}{m_2} \end{bmatrix} \tag{23}$$

with

$$m_i = 1[\mathrm{kg}], \quad k_i = 1000[\mathrm{N/m}], \quad c_i = 1[\mathrm{Ns/m}], \tag{24}$$

and control input matrix $\boldsymbol{B}$ and excitation $\boldsymbol{e}$ given by

$$\boldsymbol{B} = \begin{bmatrix} 0 \\ 0 \\ 0 \\ -\frac{1}{m_2} \end{bmatrix}, \quad \boldsymbol{e}(\boldsymbol{x}, t) = \begin{bmatrix} 0 \\ 0 \\ -\frac{F}{m_1} \\ 0 \end{bmatrix} \tag{25}$$

with

- $F(t) = \bar{F}\sin(\nu \cdot t)$,
- $\bar{F} = 5[\mathrm{N}], \quad \nu \in \ [10[1/\mathrm{s}], 80[1/\mathrm{s}]]$.

The feedback control employed is

$$\tilde{p}_k(\boldsymbol{x}, \alpha) = -u_{max} \cdot \mathrm{sat}[\tan(\alpha)\boldsymbol{x}^T\boldsymbol{P}\boldsymbol{B}\tilde{\boldsymbol{e}}_k], \quad k \in \{1, \ldots, 4\}. \tag{26}$$

with

- $u_{max} = 2.5[\mathrm{N/m}]$.

That is, we consider a fairly simple example which, however, accounts for

- mismatched uncertainty,
- constrained excitation,
- and constrained control.

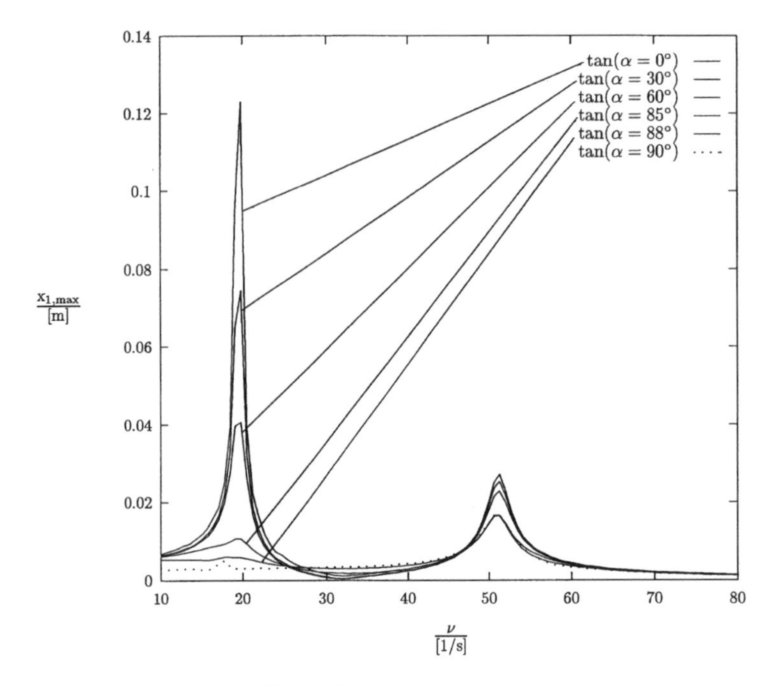

Figure 1 $x_{1,\,max}$ versus ν.

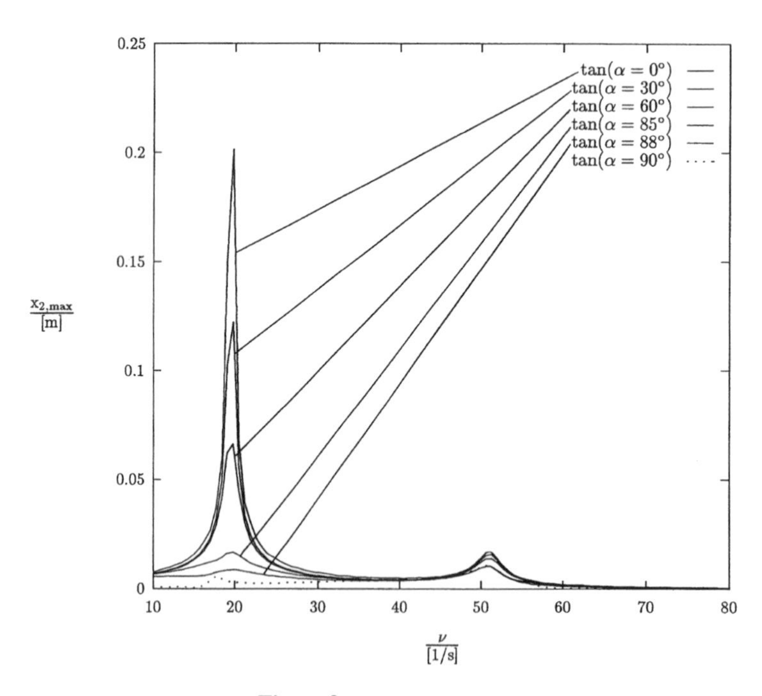

Figure 2 $x_{2,\,max}$ versus ν.

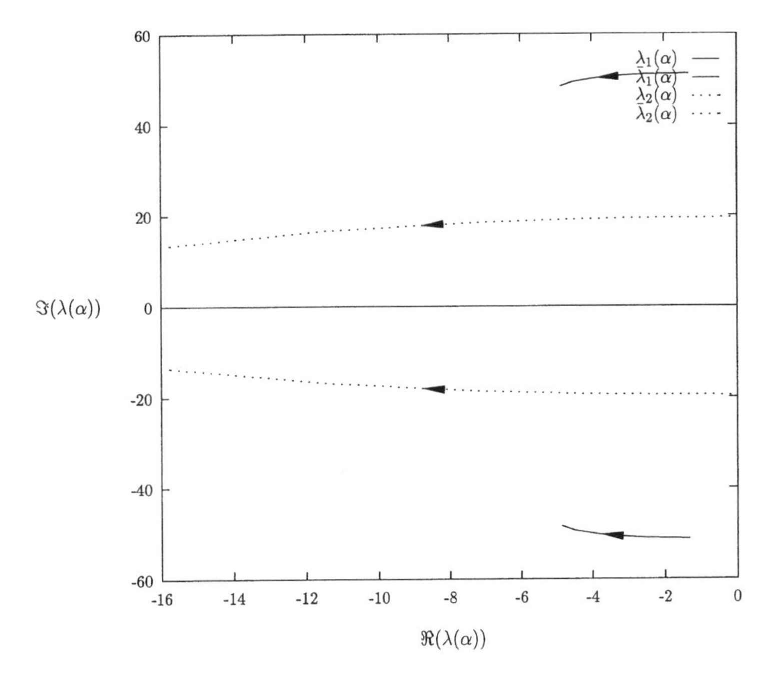

Figure 3 $\alpha \mapsto (\lambda_1, \lambda_2)$.

Figs. 1 and 2 show the amplitudes of state variables x_1 and x_2 versus the excitation frequency ν for different values of α. $\alpha = 0$ is equivalent to no control applied and $\alpha = \pi/2$ indicates the proposed Lyapunov approach. For $\alpha \to \pi/2$ the figures also show that vibration attenuation improves significantly.

Fig. 3 shows the eigenvalues of $\tilde{A}(\alpha)$. As α tends to $\pi/2$ all real parts of the eigenvalues move towards more negative values. $Re\{\lambda_1\}$ has a strong tendency towards $-\infty$, while $Re\{\lambda_2\}$ seems to move to a value less than infinity. However, on the average, the tendency towards $-\infty$ appears to hold.

References

[1] Barmish, B.R. and Leitmann, G. (1982) On ultimate boundedness control of uncertain systems in the absence of matching assumptions. *IEEE Trans. on Autom. Control*, **AC 27** (1).

[2] Chen, Y.H. (1987) On the robustness of mismatched uncertain dynamical systems. *Trans. of the ASME, Journal of Dynamical Systems, Measurement and Control*, **109** (29).

[3] Leitmann, G. (1981) On the efficacy of nonlinear control in uncertain linear systems. *J. Dyn. Syst., Measurement, and Contr.*, **102**, 95–102.

[4] Leitmann, G. and Pandey, S. (1989) A controller for a class of mismatched uncertain systems. *Proc. of the IEEE Conf. on Decision and Control.*

[5] Leitmann, G. and Reithmeier, E. (1993) An ER-material based control scheme for vibration suppression of dynamical systems with uncertain excitation. *Proc. of the MTNS 1993*, Regensburg, Germany, Aug. 2–6, vol. II, 755–760.
[6] Reithmeier, E. and Leitmann, G. (1994) Robust constrained control for vibration suppression of mismatched systems. *Proc. of the 7th Workshop on Dynamics and Control*, Ulm, Germany.
[7] Stalford H.L. (1987) Robust control of uncertain systems in the absence of matching conditions: scalar input. *Proc. of the IEEE Conf. on Decision and Control.*

3 CONTROL DESIGN FOR FLEXIBLE JOINT MANIPULATORS: MISMATCHED UNCERTAINTY AND PRACTICAL STABILITY

D.H. KIM and Y.H. CHEN*

The George W. Woodruff School of Mechanical Engineering, Georgia Institute of Technology, Atlanta, GA 30332-0405, USA

1 Introduction

We adopt the Lyapunov min–max approach for robust control design (Corless [2] and Leitmann [7]) for flexible joint manipulators. The main difficulty in directly applying the work in this area to flexible joint manipulators is the lack of the matching condition. We propose to overcome this difficulty by first introducing a state transformation. This transformation is only based on the possible bound of uncertainty and is feasible. The transformed system is *matched* and therefore the robust control can be designed straightforwardly. This in turn means that the controlled transformed system is practically stable. We then carry out a detailed analysis for the corresponding analysis of the original system. It can be shown that the original system is also practically stable.

2 Flexible Joint Manipulators

Consider an n-link serial mechanical manipulator. The links are assumed rigid. The joints are however flexible. All joints are revolute or prismatic and are directly actuated by DC-electric motors. The position of the rigid robot can be described by n generalized coordinates representing the degrees of freedom of the n joints. We denote the kinetic energy (KE) and potential energy (PE) of the robot as

$$KE = \frac{1}{2}\dot{q}^T M(q)\dot{q}, \quad PE = P(q). \tag{1}$$

*Corresponding author. E-mail: yehwa.chen@me.gatech.edu

Here q is the n-vector of joint positions, $M(q)$ is the combination of link inertia matrix and actuator inertia, i.e., $M(q) = D(q) + J$, where $D(q)$ is the link inertia matrix and J is a constant diagonal matrix representing the inertia of actuators, and $P(q)$ is the potential gravitational energy. By the Euler-Lagrange formulation we have the dynamic model (Spong and Vidyasagar [9])

$$D(q)\ddot{q} + C(q, \dot{q})\dot{q} + G(q) = u, \tag{2}$$

where $C(q,\dot{q})\dot{q}$ represents the Coriolis and centrifugal force, $G(q)$ represents the gravitational force, and u denotes the input force from the actuators.

Next, for the flexible joint robot define vectors $q_1 = [q^2\ q^4 \cdots q^{2n-2}\ q^{2n}]^T$ and $q_2 = [q^1\ q^3 \cdots q^{2n-3}\ q^{2n-1}]^T$, where q^2, $q^4 \cdots$ are link angles and q^1, $q^3 \cdots$ are joint angles. Let

$$q = \begin{bmatrix} q_1 \\ q_2 \end{bmatrix} \tag{3}$$

be the $2n$-vector of generalized coordinates for the system. We model the joint flexibility by a linear torsional spring at each joint and denote by K the diagonal matrix of joint stiffness. We assume that the rotors are modeled as uniform cylinders so that the gravitational potential energy of the system is independent of the rotor position and is therefore a function only of link position. The equation of motion is given by (Spong [8]):

$$\begin{bmatrix} D(q_1) & 0 \\ 0 & J \end{bmatrix} \begin{bmatrix} \ddot{q}_1 \\ \ddot{q}_2 \end{bmatrix} + \begin{bmatrix} C(q_1, \dot{q}_1)\dot{q}_1 \\ 0 \end{bmatrix} + \begin{bmatrix} G(q_1) \\ 0 \end{bmatrix} + \begin{bmatrix} K(q_1 - q_2) \\ -K(q_1 - q_2) \end{bmatrix} = \begin{bmatrix} 0 \\ u \end{bmatrix}, \tag{4}$$

where K is a constant diagonal matrix representing the torsional stiffness between links and joints (hence K^{-1} exists).

3 Uncertain System and Practical Stability

We consider the following class of uncertain dynamical systems:

$$\dot{\xi}(t) = f(\xi(t), \sigma(t), t) + B(\xi(t), \sigma(t), t)u(t), \tag{5}$$

where $t \in R$ is the time, $\xi(t) \in R^n$ is the state, $u(t) \in R^m$ is the control, $\sigma(t) \in R^o$ is the *uncertainty*, $f(\xi(t), \sigma(t), t)$ is the continuous system vector, and $B(\xi(t), \sigma(t), t)$ is the continuous input matrix.

Throughout, vector norms are taken to be Euclidean and matrix norms are the induced ones. In addition, we shall always use D to denote the link inertia matrix. However, a slight "abuse of notation" is indulged while the arguments of D may vary at different context.

Definition A feedback control $u(t) = p(\xi(t), t)$ renders the uncertain dynamical system (5) *practically stable* if there exists constant $\underline{d}_\xi > 0$ such that for any initial time $t_0 \in R$ and any initial state $\xi_0 \in R^n$, the following properties hold.

i) *Existence and continuation of solutions*: Given $(\xi_0, t_0) \in R^n \times R$, the closed-loop system

$$\dot{\xi}(t) = f(\xi(t), \sigma(t), t) + B(\xi(t), \sigma(t), t)p(\xi(t), t) \tag{6}$$

possesses a solution $\xi(\cdot) : (t_0, t_1) \longrightarrow R^n$, $\xi(t_0) = \xi_0$, $t_1 > t_0$. Furthermore, every solution $\xi(\cdot) : [t_0, t_1) \rightarrow R^n$ can be continued over (t_0, ∞).

ii) *Uniform boundedness*: Given any constant $r_\xi > 0$ and any solution $\xi(\cdot) : [t_0, \infty) \rightarrow R^n$, $\xi(t_0) = \xi_0$ of (6) with $\|\xi_0\| \leq r_\xi$, there exists $d_\xi(r_\xi) > 0$ such that $\|\xi(t)\| \leq d_\xi(r_\xi)$ for all $t \in [t_0, \infty)$.

iii) *Uniform ultimate boundedness*: Given any constant $\bar{d}_\xi > \underline{d}_\xi$ and any $r_\xi \in (0, \infty)$, there exists a finite time $T_\xi(\bar{d}_\xi, r_\xi)$ such that $\|\xi_0\| \leq r_\xi$ implies $\|\xi(t)\| \leq \bar{d}_\xi$ for all $t \geq t_0 + T_\xi(\bar{d}_\xi, r_\xi)$.

iv) *Uniform stability*: Given any $\bar{d}_\xi > \underline{d}_\xi$, there exists a $\delta_\xi(\bar{d}_\xi) > 0$ such that $\|\xi_0\| \leq \delta_\xi(\bar{d}_\xi)$ implies $\|\xi(t)\| \leq \bar{d}_\xi$ for all $t \geq t_0$.

Remark Uniform ultimate boundedness in practical stability means that the response of the system enters and remains within a particular neighborhood of the equilibrium position after some finite time.

4 Implanting a Robust Control

The control design procedure is carried out iteratively. Consider system (4). Let $X_1 = q_1, X_2 = \dot{q}_1, X_3 = q_2$ and $X_4 = \dot{q}_2$; also let $x_1 = [X_1^T \quad X_2^T]^T, x_2 = [X_3^T \quad X_4^T]^T$, and $x = [x_1^T \quad x_2^T]^T$. Then we construct two subsystems as follows:

$$N_1 : \dot{x}_1(t) = f_1(x_1(t), \sigma_1) + B_1(x_1(t), \sigma_1)x_2(t), \tag{7}$$

$$N_2 : \dot{x}_2(t) = f_2(x(t), \sigma_2) + B_2(\sigma_2)u(t), \tag{8}$$

where

$$f_1(x_1, \sigma_1) := \begin{bmatrix} \dot{q}_1 \\ f_{11}(x_1, \sigma_1) \end{bmatrix},$$

$$f_{11}(x_1, \sigma_1) := -D^{-1}(q_1, \sigma_1)C(q_1, \dot{q}_1, \sigma_1)\dot{q}_1 - D^{-1}(q_1, \sigma_1)G(q_1, \sigma_1) - D^{-1}(q_1, \sigma_1)K(\sigma_1)q_1,$$

$$f_2(x, \sigma_2) := \begin{bmatrix} \dot{q}_2 \\ -J^{-1}(\sigma_2)K(\sigma_2)q_2 + J^{-1}(\sigma_2)K(\sigma_2)q_1 \end{bmatrix},$$

$$B_1(x_1, \sigma_1) := \begin{bmatrix} 0 & 0 \\ D^{-1}(q_1, \sigma_1)K(\sigma_1) & 0 \end{bmatrix}, \tag{9}$$
$$B_2(\sigma_2) := \begin{bmatrix} 0 \\ J^{-1}(\sigma_2) \end{bmatrix}.$$

Here $\sigma_1 \in R_1^o$ and $\sigma_2 \in R_2^o$ are uncertain parameter vectors in N_1 and N_2, respectively.

Assumption 1 The uncertain parameters σ_1 and σ_2 are constant with $\sigma_1 \in \Sigma_1 \subset R_1^o$ and $\sigma_2 \in \Sigma_2 \subset R_2^o$ where the sets Σ_1, Σ_2 are prescribed and compact.

For a system with constant parameters we have the skew-symmetric property of the matrix $\dot{D}(q_1, \sigma_1) - 2C(q_1, \dot{q}_1, \sigma_1)$ (Spong and Vidyasagar [9]). This holds for flexible joint manipulators as well as rigid manipulators. The control problem is to design u which renders the system N_1, N_2 practically stable. The total system (7–8) does not meet the matching condition (Chen and Leitmann [1]). Thus, the control design proposed by Corless and Leitmann (1981) does not apply directly. To design robust control, we propose a two-step procedure. This is also similar to Freeman and Kokotovic [5]. Let us rewrite the first part of (4) as

$$D(q_1)\ddot{q}_1 + C(q_1, \dot{q}_1)\dot{q}_1 + G(q_1) + Kq_1 = Ku_1 + K(q_2 - u_1), \tag{10}$$

where the "control" u_1 is *implanted.* This does not affect the total dynamics in the subsystem N_1. For given $S_1 = diag[S_{1i}]_{n \times n}, S_{1i} > 0$, we choose a function $\bar{\rho}_1 : R^n \times R^n \to R_+$ such that for all $\sigma_1 \in \Sigma_1$,

$$\bar{\rho}_1(q_1, \dot{q}_1) \geq \| \phi_1(q_1, \dot{q}_1, \sigma_1) \|, \tag{11}$$

where

$$\phi_1(q_1, \dot{q}_1, \sigma_1) := -G(q_1, \sigma_1) - K(\sigma_1)q_1 + D(q_1, \sigma_1)S_1\dot{q}_1 + C(q_1, \dot{q}_1, \sigma_1)S_1 q_1. \tag{12}$$

The stiffness matrix K can be decomposed into two parts, namely, the nominal and uncertain portions:

$$K = \bar{K} + \Delta K, \tag{13}$$

where $\bar{K}$ is the known nominal portion and ΔK is the uncertain portion. Furthermore, there exists a diagonal matrix $E \in R^{n \times n}$ such that

$$\Delta K = \bar{K}\, E. \tag{14}$$

Therefore we can express $K = \bar{K}(I + E)$, where I denotes the identity matrix.

Assumption 2 There exists a constant $\lambda_E > 0$ such that

$$\lambda_E := \min_{\sigma_1 \in \Sigma_1} \{\lambda_{min}[I + \bar{K}E(\sigma_1)\bar{K}^{-1}]\} > 0. \tag{15}$$

Since $\bar{K}$ and E are diagonal matrices, (15) is satisfied if E is such that

$$e_m := \min_i \{ \min_{\sigma_1 \in \Sigma_1} (e_i(\sigma_1)) \} > -1, \quad i = 1, \cdots n, \tag{16}$$

where e_i is a diagonal component of E.

Assumption 3 The inertia matrix $D(q_1)$ is uniformly positive definite and uniformly bounded from above; that is, there exist positive scalar constants $\underline{\sigma}$ and $\bar{\sigma}$ such that

$$\underline{\sigma} I \leq D(q_1) \leq \bar{\sigma} I \quad \forall q_1 \in R^n. \tag{17}$$

Remark This assumption does not hold for arbitrary manipulators. For instance, consider the manipulator illustrated in Craig ([4], p. 209) with one revolute joint and one prismatic joint. There does not exist a finite $\bar{\sigma}$ for all $q \in R^2$. Relaxation of this assumption will be reported elsewhere.

Let the function $\rho_1 : R^{2n} \to R_+$ be chosen such that it is C^2 (i.e., 2-times continuously differentiable) and

$$\rho_1(x_1) \geq (1 + e_m)^{-1} \bar{\rho}_1(x_1). \tag{18}$$

For given scalar $\epsilon_1 > 0$, choose the implanted control u_1 such that

$$\begin{aligned} u_1(t) = \bar{K}^{-1} &(-K_{p1} q_1(t) - K_{v1} \dot{q}_1(t) + p_1(x_1(t), \rho_1(x_1(t)), \epsilon_1)) \\ &- \bar{K}^{-1} \beta_1 (\dot{q}_1(t) + S_1 q_1(t)), \end{aligned} \tag{19}$$

where

$$\begin{aligned} K_{p1} &:= diag[k_{p1i}]_{n \times n}, \; k_{p1i} > 0, \\ K_{v1} &:= diag[k_{v1i}]_{n \times n}, \; k_{v1i} > 0, \\ \mu_1(x_1) &= (\dot{q}_1 + S_1 q_1) \rho_1(x_1), \\ \mu_1 &= [\mu_{11} \; \mu_{12} \cdots \mu_{1n}]^T, \\ p_1 &= [p_{11} \; p_{12} \cdots p_{1n}]^T. \end{aligned} \tag{20}$$

p_{1i} is given by

$$p_{1i} = \begin{cases} -\frac{\mu_{1i}(x_1)}{\|\mu_{1i}(x_1)\|} \rho_1(x_1), & \text{if } \|\mu_{1i}(x_1)\| > \epsilon_1 \\ -\sin(\frac{\pi \mu_{1i}(x_1)}{2\epsilon_1}) \rho_1(x_1), & \text{if } \|\mu_{1i}(x_1)\| \leq \epsilon_1 \end{cases}, \tag{21}$$

$i = 1, 2, \ldots n$. Note that

$$p_{1i} \begin{cases} \leq -\frac{\mu_{1i}(x_1)}{\epsilon_1} \rho_1(x_1), & \text{if } 0 \leq \mu_{1i}(x_1) \leq \epsilon_1 \\ \geq -\frac{\mu_{1i}(x_1)}{\epsilon_1} \rho_1(x_1), & \text{if } -\epsilon_1 \leq \mu_{1i}(x_1) \leq 0 \end{cases}, \tag{22}$$

and $\|p_{1i}\| \leq \rho_1$. Such a control scheme is also seen in Han and Chen [6]. Also, $\beta_1 > 0$ is a scalar design parameter. The choice of β_1 will be made later.

Remark The control $u_1 = p_1$, with p_{1i} given by (21), is twice differentiable. This is due to that $\rho_1(\cdot)$ is C^2 and p_{1i} is constructed according to (21). The transition of the control during $\|\mu_{1i}(x_1)\| = \epsilon_1$ is smooth (by the choice of the sin function in (21)). This in fact also explains the choice of (21), which is different from Corless and Leitmann [3].

5 State Transformation

We now transform the system (N_1, N_2) to a system $(\hat{N}_1, \hat{N}_2)$ as follows. First let $z_1 = [Z_1^T \quad Z_2^T]^T$, $z_2 = [Z_3^T \quad Z_4^T]^T$, and $z = [z_1^T \quad z_2^T]^T$, where

$$\begin{aligned} Z_1 &:= q_1, \\ Z_2 &:= \dot{q}_1, \\ Z_3 &:= q_2 - u_1, \\ Z_4 &:= \dot{q}_2 - \dot{u}_1. \end{aligned} \tag{23}$$

This implies that $z_1 = x_1$ and $z_2 = x_2 - [u_1 \quad \dot{u}_1]^T$. From now on, we shall omit the uncertainty argument in $D(q_1, \sigma_1)$, $C(q_1, \dot{q}_1, \sigma_1)$, and $K(\sigma_1)$ etc. when no confusions are likely to arise. Next, the dynamics of the manipulator are expressed in terms of z:

$$\hat{N}_1 : D(Z_1)\ddot{Z}_1 = -C(Z_1, \dot{Z}_1)\dot{Z}_1 - G(Z_1) - KZ_1 + KZ_3 + Ku_1, \tag{24}$$

$$\hat{N}_2 : J\ddot{Z}_3 = -J\ddot{u}_1 - KZ_3 + KZ_1 - Ku_1 + u. \tag{25}$$

The dynamics of $\hat{N}_1$ contain the implanted control u_1. The dynamics of $\hat{N}_2$ contain, besides u, u_1 and $\ddot{u}_1$.

We now propose the control design. For given $S_2 = diag[S_{2i}]_{n\times n}$, $S_{2i} > 0$, we choose a known scalar function $\rho_2 : R^{2n} \times R^n \times R^n \to R_+$ such that

$$\rho_2(x_1, Z_3, Z_4) \geq \|\phi_2(x_1, Z_3, Z_4, \sigma_1, \sigma_2)\|, \tag{26}$$

where

$$\begin{aligned} \phi_2(x_1, Z_3, Z_4, \sigma_1, \sigma_2) :=& - J(\sigma_2)\ddot{u}_1(x_1, Z_3, Z_4, \sigma_1, \sigma_2) - K(\sigma_2)Z_3 + K(\sigma_2)Z_1 \\ & - K(\sigma_2)u_1(x_1) + J(\sigma_2)S_2\dot{Z}_3. \end{aligned} \tag{27}$$

For given scalar $\epsilon_2 > 0$, choose the control input u as follows:

$$u(t) = -K_{p2}Z_3(t) - K_{v2}\dot{Z}_3(t) + p_2(x_1(t), Z_3(t), Z_4(t), \epsilon_2), \tag{28}$$

where

$$p_2(x_1, Z_3, Z_4, \epsilon_2) = \begin{cases} -\frac{\mu_2(z_1,z_2)}{\|\mu_2(z_1,z_2)\|}\rho_2(x_1, Z_3, Z_4) & \text{if } \|\mu_2(z_1, z_2)\| > \epsilon_2, \\ -\frac{\mu_2(z_1,z_2)}{\epsilon_2}\rho_2(x_1, Z_3, Z_4) & \text{if } \|\mu_2(z_1, z_2)\| \leq \epsilon_2, \end{cases} \tag{29}$$

$$\mu_2(z_1, z_2) = (\dot{Z}_3 + S_2 Z_3)\rho_2(x_1, Z_3, Z_4), \tag{30}$$

$$K_{p2} := diag[k_{p2i}]_{n\times n}, \quad k_{p2i} > 0, \tag{31}$$

$$K_{v2} := diag[k_{v2i}]_{n\times n}, \quad k_{v2i} > 0. \tag{32}$$

The selection of β_1 in (19) for u_1 is shown below:

Let

$$\underline{\lambda}_2 := \min\{\lambda_{\min}(K_{v2}), \lambda_{\min}(S_2 K_{p2})\}. \tag{33}$$

Step 1: Select $\underline{\lambda}_2$ and compute a proper value for λ_k where $\max_{\sigma_1 \in \Sigma_1} \|K\| \le \lambda_k$ and compute λ_E in (15).

Step 2: Choose w_1 such that $w_1 > \lambda_k / 2\underline{\lambda}_2$.

Step 3: By w_1 obtained from Step 2 choose β_1 such that $\beta_1 > w_1 \lambda_k / 2\lambda_E$.

Theorem 1 *Subject to Assumptions 1–3, the system (24)–(25) is practically stable under the control (28). Furthermore, the size of the uniform ultimate boundedness region can be made arbitrary small by a suitable choice of ϵ_1 and ϵ_2.*

Proof

Choose the Lyapunov function candidate

$$V(z_1, z_2) = V_1(z_1) + V_2(z_2), \tag{34}$$

where

$$V_1(z_1) = \frac{1}{2}(Z_2 + S_1 Z_1)^T D (Z_2 + S_1 Z_1) + \frac{1}{2} Z_1^T (\bar{K}_{p1} + S_1 \bar{K}_{v1}) Z_1, \tag{35}$$

$$V_2(z_2) = \frac{1}{2}(Z_4 + S_2 Z_3)^T J (Z_4 + S_2 Z_3) + \frac{1}{2} Z_3^T (K_{p2} + S_2 K_{v2}) Z_3, \tag{36}$$

and

$$\begin{aligned} \bar{K}_{p1} &= K_{p1} + \bar{K} E \bar{K}^{-1} K_{p1}, \\ \bar{K}_{v1} &= K_{v1} + \bar{K} E \bar{K}^{-1} K_{v1}. \end{aligned} \tag{37}$$

From (37) we see that $\bar{K}_{p1}$ and $\bar{K}_{v1}$ are positive definite. To show that V is indeed legitimate, we shall prove that both V_1 and V_2 are positive definite and decrescent. Based on Assumption 3,

$$\begin{aligned} V_1 &\ge \frac{1}{2}\underline{\sigma}\|Z_2 + S_1 Z_1\|^2 + \frac{1}{2} Z_1^T (\bar{K}_{p1} + S_1 \bar{K}_{v1}) Z_1 \\ &= \frac{1}{2}\underline{\sigma}\sum_{i=1}^{n} (\dot{Z}_{1i}^2 + 2 S_{1i} \dot{Z}_{1i} Z_{1i} + S_{1i}^2 Z_{1i}^2) + \frac{1}{2}\sum_{i=1}^{n} (\bar{k}_{p1i} + S_{1i} \bar{k}_{v1i}) Z_{1i}^2 \\ &= \frac{1}{2}\sum_{i=1}^{n} [Z_{1i} \; \dot{Z}_{1i}] \underline{\Omega}_{1i} \begin{bmatrix} Z_{1i} \\ Z_{2i} \end{bmatrix}, \end{aligned} \tag{38}$$

where

$$\underline{\Omega}_{1i} := \begin{bmatrix} \underline{\sigma}S_{1i}^2 + \bar{k}_{p1i} + S_{1i}\bar{k}_{v1i} & \underline{\sigma}S_{1i} \\ \underline{\sigma}S_{1i} & \underline{\sigma} \end{bmatrix}, \tag{39}$$

where Z_{1i} and $\dot{Z}_{1i}$ are the i-th components of Z_1 and $\dot{Z}_1$, respectively. Since $\underline{\Omega}_{1i} > 0 \ \forall i$, V_1 is positive definite:

$$\begin{aligned} V_1 &\geq \frac{1}{2}\sum_{i=1}^{n} \lambda_{\min}(\underline{\Omega}_{1i})(Z_{1i}^2 + \dot{Z}_{1i}^2) \\ &\geq \gamma_1^1 \|z_1\|^2, \end{aligned} \tag{40}$$

where

$$\gamma_1^1 := \frac{1}{2}\min_i\{\min_{\sigma_1 \in \Sigma_1} \lambda_{\min}(\underline{\Omega}_{1i}), i = 1, 2, \cdots, n\}. \tag{41}$$

Next,

$$\begin{aligned} V_1 &\leq \frac{1}{2}\bar{\sigma}\|Z_2 + S_1 Z_1\|^2 + \frac{1}{2}Z_1^T(\bar{K}_{p1} + S_1\bar{K}_{v1})Z_1 \\ &= \frac{1}{2}\bar{\sigma}\sum_{i=1}^{n}(\dot{Z}_{1i}^2 + 2S_{1i}\dot{Z}_{1i}Z_{1i} + S_{1i}^2 Z_{1i}^2) + \frac{1}{2}\sum_{i=1}^{n}(\bar{k}_{p1i} + S_{1i}\bar{k}_{v1i})Z_{1i}^2 \\ &= \frac{1}{2}\sum_{i=1}^{n}[Z_{1i}\ \dot{Z}_{1i}]\bar{\Omega}_{1i}\begin{bmatrix} Z_{1i} \\ \dot{Z}_{1i} \end{bmatrix}, \end{aligned} \tag{42}$$

where

$$\bar{\Omega}_{1i} := \begin{bmatrix} \bar{\sigma}S_{1i}^2 + \bar{k}_{p1i} + S_{1i}\bar{k}_{v1i} & \bar{\sigma}S_{1i} \\ \bar{\sigma}S_{1i} & \bar{\sigma} \end{bmatrix}. \tag{43}$$

Therefore, we have

$$\begin{aligned} V_1 &\leq \frac{1}{2}\sum_{i=1}^{n} \lambda_{\max}(\bar{\Omega}_{1i})(Z_{1i}^2 + \dot{Z}_{1i}^2) \\ &\leq \gamma_2^1 \|z_1\|^2, \end{aligned} \tag{44}$$

where

$$\gamma_2^1 := \frac{1}{2}\max_i\left\{\max_{\sigma_1 \in \Sigma_1} \lambda_{\max}(\bar{\Omega}_{1i}), \quad i = 1, 2, \cdots, n\right\}. \tag{45}$$

Similar to V_1, V_2 is also positive definite and decrescent. This is so since

$$\frac{1}{2}\sum_{i=1}^{n} \lambda_{\min}(\underline{\Omega}_{2i})(Z_{3i}^2 + \dot{Z}_{3i}^2) \leq V_2 \leq \frac{1}{2}\sum_{i=1}^{n} \lambda_{\max}(\bar{\Omega}_{2i})(Z_{3i}^2 + \dot{Z}_{3i}^2) \tag{46}$$

and therefore

$$\gamma_1^2\|z_2\|^2 \leq V_2(z_2) \leq \gamma_2^2\|z_2\|^2, \tag{47}$$

where

$$
\begin{aligned}
\gamma_1^2 &:= \frac{1}{2}\min_i\{\lambda_{\min}(\underline{\Omega}_{2i}), \quad i = 1, 2, \cdots, n\}, \\
\gamma_2^2 &:= \frac{1}{2}\max_i\{\lambda_{\max}(\bar{\Omega}_{2i}), \quad i = 1, 2, \cdots, n\}, \\
\underline{\Omega}_{2i} &:= \begin{bmatrix} \underline{\theta} S_{2i}^2 + k_{p2i} + S_{2i}k_{v2i} & \underline{\theta} S_{2i} \\ \underline{\theta} S_{2i} & \underline{\theta} \end{bmatrix}, \\
\bar{\Omega}_{2i} &:= \begin{bmatrix} \bar{\theta} S_{2i}^2 + k_{p2i} + S_{2i}k_{v2i} & \bar{\theta} S_{2i} \\ \bar{\theta} S_{2i} & \bar{\theta} \end{bmatrix}, \\
\bar{\theta} &:= \lambda_{\max}(J), \\
\underline{\theta} &:= \lambda_{\min}(J).
\end{aligned}
\tag{48}
$$

The derivative of V_1 along a trajectory of the controlled system (24)–(25) is given by

$$
\begin{aligned}
\dot{V}_1 = {} & (\dot{Z}_1 + S_1 Z_1)^T D(\ddot{Z}_1 + S_1\dot{Z}_1) + \frac{1}{2}(\dot{Z}_1 + S_1 Z_1)^T \dot{D}(\dot{Z}_1 + S_1 Z_1) \\
& + Z_1^T(\bar{K}_{p1} + S_1\bar{K}_{v1})\dot{Z}_1.
\end{aligned}
\tag{50}
$$

By the skew-symmetric property of $\dot{D} - 2C$, it can be shown that

$$
\begin{aligned}
\dot{V}_1 = {} & (\dot{Z}_1 + S_1 Z_1)^T(-G - KZ_1 + DS_1\dot{Z}_1 + CS_1 Z_1 + Ku_1 + KZ_3) \\
& + Z_1^T(\bar{K}_{p1} + S_1\bar{K}_{v1})\dot{Z}_1.
\end{aligned}
\tag{51}
$$

According to (11), it can be seen that

$$
\begin{aligned}
\dot{V}_1 = {} & (\dot{Z}_1 + S_1 Z_1)^T(\phi_1 + \bar{K}u_1 + \bar{K}Eu_1) \\
& + Z_1^T(\bar{K}_{p1} + S_1\bar{K}_{v1})\dot{Z}_1 + (\dot{Z}_1 + S_1 Z_1)^T KZ_3.
\end{aligned}
\tag{52}
$$

It follows from (19)

$$
\begin{aligned}
\dot{V}_1 = {} & (\dot{Z}_1 + S_1 Z_1)^T\phi_1 + (\dot{Z}_1 + S_1 Z_1)^T\big(-K_{p1}Z_1 - K_{v1}\dot{Z}_1 + p_1 - \beta_1(\dot{Z}_1 + S_1 Z_1)\big) \\
& + (\dot{Z}_1 + S_1 Z_1)^T\bar{K}E\bar{K}^{-1}\big(-K_{p1}Z_1 - K_{v1}\dot{Z}_1 + p_1 - \beta_1(\dot{Z}_1 + S_1 Z_1)\big) \\
& + Z_1^T(\bar{K}_{p1} + S_1\bar{K}_{v1})\dot{Z}_1 + (\dot{Z}_1 + S_1 Z_1)^T KZ_3 \\
\leq {} & - \dot{Z}_1^T\bar{K}_{v1}\dot{Z}_1 - Z_1^T S_1\bar{K}_{p1}Z_1 + (\dot{Z}_1 + S_1 Z_1)^T\phi_1 \\
& + (\dot{Z}_1 + S_1 Z_1)^T p_1 + (\dot{Z}_1 + S_1 Z_1)^T\bar{K}E\bar{K}^{-1}p_1 \\
& - \beta_1\lambda_E\|\dot{Z}_1 + S_1 Z_1\|^2 + (\dot{Z}_1 + S_1 Z_1)^T KZ_3 \\
\leq {} & - \underline{\lambda}_1\|z_1\|^2 - \beta_1\lambda_E\|\dot{Z}_1 + S_1 Z_1\|^2 + (\dot{Z}_1 + S_1 Z_1)^T\phi_1 \\
& + (\dot{Z}_1 + S_1 Z_1)^T p_1 + (\dot{Z}_1 + S_1 Z_1)^T\bar{K}E\bar{K}^{-1}p_1 + (\dot{Z}_1 + S_1 Z_1)^T KZ_3,
\end{aligned}
\tag{53}
$$

where

$$
\underline{\lambda}_1 = \lambda_E\min\{\lambda_{\min}(K_{v1}), \lambda_{\min}(S_1K_{p1})\}. \tag{54}
$$

If $\|\mu_{1i}\| > \epsilon_1$,

$$
\begin{aligned}
&(\dot{Z}_1 + S_1 Z_1)^T \phi_1 + (\dot{Z}_1 + S_1 Z_1)^T p_1 + (\dot{Z}_1 + S_1 Z_1)^T \bar{K} E \bar{K}^{-1} p_1 \\
&\leq \sum_{i=1}^{n} \|\dot{Z}_{1i} + S_{1i} Z_{1i}\| \bar{\rho}_1 + \sum_{i=1}^{n} (\dot{Z}_{1i} + S_{1i} Z_{1i}) p_{1i} + \sum_{i=1}^{n} (\dot{Z}_{1i} + S_{1i} Z_{1i}) e_i p_{1i} \\
&\leq \sum_{i=1}^{n} \|\dot{Z}_{1i} + S_{1i} Z_{1i}\| (1 + e_m) \rho_1 + \sum_{i=1}^{n} (\dot{Z}_{1i} + S_{1i} Z_{1i}) \left(-\frac{\dot{Z}_{1i} + S_{1i} Z_{1i}}{\|\dot{Z}_{1i} + S_{1i} Z_{1i}\|} \rho_1 \right) \\
&+ \sum_{i=1}^{n} (\dot{Z}_{1i} + S_{1i} Z_{1i}) e_i \left(-\frac{\dot{Z}_{1i} + S_{1i} Z_{1i}}{\|\dot{Z}_{1i} + S_{1i} Z_{1i}\|} \rho_1 \right) \\
&\leq \sum_{i=1}^{n} \|\dot{Z}_{1i} + S_{1i} Z_{1i}\| (1 + e_m) \rho_1 - \sum_{i=1}^{n} \|\dot{Z}_{1i} + S_{1i} Z_{1i}\| \rho_1 - e_m \sum_{i=1}^{n} \|\dot{Z}_{1i} + S_{1i} Z_{1i}\| \rho_1 \\
&= 0.
\end{aligned}
\tag{55}
$$

If $\|\mu_{1i}\| \leq \epsilon_1$,

$$
\begin{aligned}
&(\dot{Z}_1 + S_1 Z_1)^T \phi_1 + (\dot{Z}_1 + S_1 Z_1)^T p_1 + (\dot{Z}_1 + S_1 Z_1)^T \bar{K} E \bar{K}^{-1} p_1 \\
&\leq \sum_{i=1}^{n} \|\dot{Z}_{1i} + S_{1i} Z_{1i}\| \bar{\rho}_1 + \sum_{i=1}^{n} \|\dot{Z}_{1i} + S_{1i} Z_{i1}\|^2 \left(-\frac{\rho_1^2}{\epsilon_1} \right) - e_m \sum_{i=1}^{n} \|\dot{Z}_1 + S_{1i} Z_{1i}\|^2 \frac{\rho_1^2}{\epsilon_1} \\
&= \sum_{i=1}^{n} (1 + e_m) \rho_1 - (1 + e_m) \sum_{i=1}^{n} \|\dot{Z}_{1i} + S_{1i} Z_{1i}\|^2 \frac{\rho_1^2}{\epsilon_1} \\
&\leq \frac{n(1 + e_m)\epsilon_1}{4}
\end{aligned}
\tag{56}
$$

Based on inequalities $ab \leq \frac{1}{2}(a^2 + b^2)$, $a, b \in R$, and $\|Z_3\|^2 \leq \|z_2\|^2$, we have that for any constant $w_1 > 0$,

$$
\begin{aligned}
(\dot{Z}_1 + S_1 Z_1)^T K Z_3 &\leq \|\dot{Z}_1 + S_1 Z_1\| \|Z_3\| \|K\| \\
&\leq \left(\frac{1}{2} w_1 \|\dot{Z}_1 + S_1 Z_1\|^2 + \frac{1}{2} w_1^{-1} \|Z_3\|^2 \right) \|K\| \\
&\leq \left(\frac{1}{2} w_1 \|\dot{Z}_1 + S_1 Z_1\|^2 + \frac{1}{2} w_1^{-1} \|z_2\|^2 \right) \|K\|. \\
&\leq \left(\frac{1}{2} w_1 \|\dot{Z}_1 + S_1 Z_1\|^2 + \frac{1}{2} w_1^{-1} \|z_2\|^2 \right) \lambda_k,
\end{aligned}
\tag{57}
$$

This leads to

$$
\begin{aligned}
\dot{V}_1 \leq &-\underline{\lambda}_1 \|z_1\|^2 + \frac{n(1 + e_m)\epsilon_1}{4} - \beta_1 (1 + \lambda_E) \|\dot{Z}_1 + S_1 Z_1\|^2 \\
&+ \left(\frac{1}{2} w_1 \|\dot{Z}_1 + S_1 Z_1\|^2 + \frac{1}{2} w_1^{-1} \|z_2\|^2 \right) \lambda_k.
\end{aligned}
\tag{58}
$$

Next, the derivative of $V_2(Z_2)$ along the trajectory of (24)–(25) is given by

$$\begin{aligned}\dot{V}_2 &= (\dot{Z}_3 + S_2Z_3)^T J(\ddot{Z}_3 + S_2\dot{Z}_3) + Z_3^T(K_{p2} + S_2K_{v2})\dot{Z}_3 \\ &= (\dot{Z}_3 + S_2Z_3)^T(-J\ddot{u}_1 - KZ_3 + KZ_1 - Ku_1 + u + JS_2\dot{Z}_3) \\ &\quad + Z_3^T(K_{p2} + S_2K_{v2})\dot{Z}_3.\end{aligned} \tag{59}$$

It follows from (27), (28), and (33) that

$$\begin{aligned}\dot{V}_2 &= (\dot{Z}_3 + S_2Z_3)^T(\phi_2 + u) + Z_3^T(K_{p2} + S_2K_{v2})\dot{Z}_3 \\ &\leq (\dot{Z}_3 + S_2Z_3)^T p_2 + (\dot{Z}_3 + S_2Z_3)^T\phi_2 - \underline{\lambda}_2\|z_2\|^2 \\ &\leq (\dot{Z}_3 + S_2Z_3)^T p_2 + \|\dot{Z}_3 + S_2Z_3\|\rho_2 - \underline{\lambda}_2\|z_2\|^2.\end{aligned} \tag{60}$$

If $\|\mu_2\| > \epsilon_2$,

$$\begin{aligned}&(\dot{Z}_3 + S_2Z_3)^T p_2 + \|\dot{Z}_3 + S_2Z_3\|\rho_2 \\ &\leq (\dot{Z}_3 + S_2Z_3)^T\left(-\frac{\mu_2}{\|\mu_2\|}\rho_2\right) + \|\dot{Z}_3 + S_2Z_3\|\rho_2 \\ &\leq -\|\mu_2\| + \|\mu_2\| \\ &= 0.\end{aligned} \tag{61}$$

If $\|\mu_2\| \leq \epsilon_2$,

$$\begin{aligned}&(\dot{Z}_3 + S_2Z_3)^T p_2 + \|\dot{Z}_3 + S_2Z_3\|\rho_2 \\ &= (\dot{Z}_3 + S_2Z_3)^T\left(-\frac{\dot{Z}_3 + S_2Z_3}{\epsilon_2}\rho_2^2\right) + \|\dot{Z}_3 + S_2Z_3\|\rho_2 \\ &\leq -\frac{\|\mu_2\|^2}{\epsilon_2} + \|\mu_2\| \\ &\leq \frac{\epsilon_2}{4}.\end{aligned} \tag{62}$$

Thus, $\dot{V}_2$ is upper bounded:

$$\dot{V}_2 \leq -\underline{\lambda}_2\|z_2\|^2 + \frac{\epsilon_2}{4}. \tag{63}$$

Now, using (58) and (63),

$$\begin{aligned}\dot{V} &= \dot{V}_1 + \dot{V}_2 \\ &\leq -\underline{\lambda}_1\|z_1\|^2 + \frac{n(1+e_m)\epsilon_1}{4} - \beta_1\lambda_E\|\dot{Z}_1 + S_1Z_1\|^2 \\ &\quad + \left(\frac{1}{2}w_1\|\dot{Z}_1 + S_1Z_1\|^2 + \frac{1}{2}w_1^{-1}\|z_2\|^2\right)\lambda_k - \underline{\lambda}_2\|z_2\|^2 + \frac{\epsilon_2}{4} \\ &= -\underline{\lambda}_1\|z_1\|^2 - \left(\beta_1\lambda_E - \frac{1}{2}w_1\lambda_k\right)\|\dot{Z}_1 + S_1Z_1\|^2 \\ &\quad - \left(\underline{\lambda}_2 - \frac{1}{2}w_1^{-1}\lambda_k\right)\|z_2\|^2 + \frac{n(1+e_m)\epsilon_1}{4} + \frac{\epsilon_2}{4}.\end{aligned} \tag{64}$$

If $\underline{\lambda}_2 - \frac{1}{2}w_1^{-1}\lambda_k > 0$ and $\beta_1\lambda_E - \frac{1}{2}w_1\lambda_k > 0$, then we have

$$\dot{V} \leq -\min\left\{\underline{\lambda}_1, \underline{\lambda}_2 - \frac{1}{2}w_1^{-1}\lambda_k\right\}\|z\|^2 + \bar{k}, \tag{65}$$

where $\bar{k} := \frac{n(1+e_m)\epsilon_1}{4} + \frac{\epsilon_2}{4}$. Following (65) for $r_z \geq 0$, if $\|z_0\| \leq r_z$, we can satisfy the requirements of uniform boundedness, uniform ultimate boundedness and uniform stability by selecting (Corless and Leitmann [3])

$$d_z(r_z) = \begin{cases} R_z\sqrt{\frac{\gamma_2}{\gamma_1}} & \text{if } r_z \leq R_z \\ r_z\sqrt{\frac{\gamma_2}{\gamma_1}} & \text{if } r_z > R_z, \end{cases} \tag{66}$$

$$T_z(\bar{d}_z, r_z) = \begin{cases} 0 & \text{if } r_z \leq \bar{d}_z\sqrt{\frac{\gamma_1}{\gamma_2}} \\ \frac{\gamma_2 r_z^2 - \gamma_1^2\gamma_2^{-1}\bar{d}_z^2}{\gamma_1\gamma_2^{-1}\gamma_3\bar{d}_z^2 - \bar{k}} & \text{otherwise}, \end{cases} \tag{67}$$

$$\delta_z(\bar{d}_z) = R_z, \tag{68}$$

where $\gamma_1 = \min\{\gamma_1^1, \gamma_1^2\}$, $\gamma_2 = \max\{\gamma_2^1, \gamma_2^2\}$, $R_z = \sqrt{\frac{\bar{k}}{\gamma_3}}$, $\gamma_3 = \min\{\underline{\lambda}_1, \underline{\lambda}_2 - \frac{1}{2}w_1^{-1}\lambda_k\}$, $\underline{d}_z = R_z\sqrt{\frac{\gamma_2}{\gamma_1}}$. The constant $\underline{d}_z$, which determines the size of the uniform ultimate boundedness ball and uniform stability ball, approaches 0 as R_z approaches 0. It can be seen that $R_z \to 0$ as both ϵ_1 and $\epsilon_2 \to 0$. Therefore if both ϵ_1 and $\epsilon_2 \to 0$, then $\underline{d}_z$ converges to 0. Q.E.D.

6 System with Time-varying Uncertainty

We have considered the system with constant uncertainty in Section 5. When the parameters are time-varying (which may be due to, e.g., time-varying payload, inertia, and stiffness in joints), $\dot{D} - 2C$ is no longer skew symmetric. Therefore the implanted robust control needs to be modified.

Assumption 4 The mappings $\sigma_1(\cdot) : R \to \Sigma_1 \subset R^{o_1}$, $\sigma_2(\cdot) : R \to \Sigma_2 \subset R^{o_2}$ are Lebesgue measurable with Σ_1, Σ_2 prescribed and compact. Furthermore, the mappings $\dot{\sigma}_1(\cdot) : R \to \Sigma_{1t} \subset R^{o_1}$, $\dot{\sigma}_2(\cdot) : R \to \Sigma_{2t} \subset R^{o_2}$ are Lebesgue measurable with Σ_{1t}, Σ_{2t} prescribed and compact.

Subject to Assumption 4, the $\phi_1(\cdot)$ in (12) and $\phi_2(\cdot)$ in (27) are changed to

$$\begin{aligned}\phi_1(q_1, \dot{q}_1, \sigma_1(t)) := \frac{1}{2}\dot{D}(q_1, \dot{q}_1, \sigma_1(t))(\dot{q}_1 + S_1q_1) - C(q_1, \dot{q}_1, \sigma_1(t))\dot{q}_1 \\ - G(q_1, \sigma_1(t)) - K(\sigma_1(t))q_1 + D(q_1, \sigma_1(t))S_1\dot{q}_1,\end{aligned} \tag{69}$$

$$\begin{aligned}
&\phi_2\big(x_1, z_2, \sigma_1(t), \sigma_2(t)\big) \\
&\quad := -J\big(\sigma_2(t)\big)\ddot{u}_1\big(x_1, z_2, \sigma_1(t), \sigma_2(t)\big) - K\big(\sigma_2(t)\big)Z_3 + K\big(\sigma_2(t)\big)Z_1 \\
&\quad - K\big(\sigma_2(t)\big)u_1(x_1) + J\big(\sigma_2(t)\big)S_2\dot{Z}_3 + \frac{1}{2}\dot{J}\big(\sigma_2(t)\big)(\dot{Z}_3 + S_2 Z_3),
\end{aligned} \tag{70}$$

This in turn requires different choices of the bounding functions $\bar{\rho}_1(q_1, \dot{q}_1)$ and $\rho_2(x_1, z_2)$ (which bound $\phi_{1,2}(\cdot)$ in the same way as was shown in (12) and (26)). Based on the different bounding functions we use the procedure described in Sections 3 and 4 to obtain control of the $u(\cdot)$. We can also see the difference in (50) and (59) for the system with time-varying uncertainty when proving Theorem 1. As there is no longer the skew-symmetric property of $\dot{D} - 2C$, we need the modified $\phi_1(\cdot)$ as (69) and $\phi_2(\cdot)$ as (70) to prove Theorem 1.

7 Performance Analysis of the System (N_1, N_2)

Since the original system is described in terms of x rather than z, it is necessary to investigate the performance of the original system when the transformed system is practically stable. Suppose z is bounded by a constant δ:

$$\|z\| \leq \delta. \tag{71}$$

Then

$$\begin{aligned}
\|z\|^2 &= \|z_1\|^2 + \|z_2\|^2 \\
&\leq \delta^2.
\end{aligned} \tag{72}$$

This implies that, by (23),

$$\|x_1\| \leq \delta, \ \|x_2 - [u_1 \ \dot{u}_1]^T\| \leq \delta. \tag{73}$$

From (23) we have

$$x_2 = z_2 + \begin{bmatrix} u_1 \\ \dot{u}_1 \end{bmatrix}. \tag{74}$$

The control u_1 is bounded by

$$\begin{aligned}
\|u_1\| &\leq \|\bar{K}^{-1}\|\big(\|K_{p1}\|\|Z_1\| + \|K_{v1}\|\|\dot{Z}_1\| + \|p_1\| + \beta_1\|\dot{Z}_1 + S_1 Z_1\|\big) \\
&\leq a_1\|Z_1\| + a_2\|Z_2\| + \|p_1\| \\
&\leq \max(a_1, a_2)\sqrt{2}\|z_1\| + \|p_1\| \\
&=: a_3\|z_1\| + \|p_1\|
\end{aligned} \tag{75}$$

where

$$\begin{aligned} a_1 &:= \|\bar{K}^{-1}\|(\|K_{p1}\| + \beta_1\|S_1\|), \\ a_2 &:= \|\bar{K}^{-1}\|(\|K_{v1}\| + \beta_1), \\ a_3 &:= \max(a_1, a_2)\sqrt{2}. \end{aligned} \tag{76}$$

From (21) we get

$$\begin{aligned} \|p_1\| &\leq \|p_{11}\| + \|p_{12}\| + \cdots \|p_{1n}\| \\ &\leq \sqrt{n}\rho_1(x_1). \end{aligned} \tag{77}$$

Furthermore, by the continuity property of $\rho_1(\cdot)$, if $\|x_1\| \leq \delta$ there exists a constant $\hat{p}_1(\delta)$ such that

$$\|u_1\| \leq a_3\delta + \hat{p}_1(\delta). \tag{78}$$

Next, with continuity property of $\dot{p}_1(\cdot)$, there exist constants $\bar{\eta}_1, \bar{\eta}_2$ such that

$$\begin{aligned} \|\dot{p}_1\| &= \left\|\frac{\partial p_1}{\partial Z_1}\dot{Z}_1\right\| + \left\|\frac{\partial p_1}{\partial Z_2}\dot{Z}_2\right\| \\ &\leq \sup_{\|z_1\|\leq\delta}\left\|\frac{\partial p_1}{\partial Z_1}\right\|\|\dot{Z}_1\| + \sup_{\|z_1\|\leq\delta}\left\|\frac{\partial p_1}{\partial Z_2}\right\|\|\dot{Z}_2\| \\ &=: \bar{\eta}_1\|\dot{Z}_1\| + \bar{\eta}_2\|\dot{Z}_2\|, \end{aligned} \tag{79}$$

where

$$\begin{aligned} \bar{\eta}_1 &:= \sup_{\|z_1\|\leq\delta}\left\|\frac{\partial p_1}{\partial Z_1}\right\| < \infty, \\ \bar{\eta}_2 &:= \sup_{\|z_1\|\leq\delta}\left\|\frac{\partial p_1}{\partial Z_2}\right\| < \infty. \end{aligned} \tag{80}$$

From (24) we have

$$\begin{aligned} \|\ddot{Z}_1\| &\leq \|D^{-1}C\|\|\dot{Z}_1\| + \|D^{-1}G\| + \|D^{-1}K\|(\|Z_1\| + \|u_1\| + \|Z_3\|) \\ &\leq \sup_{\|z_1\|\leq\delta}\|D^{-1}C\|\delta + \sup_{\|Z_1\|\leq\delta}\|D^{-1}G\| + \sup_{\|Z_1\|\leq\delta}\|D^{-1}K\|(\delta + a_3\delta + \hat{p}_1(\delta) + \delta) \\ &=: h_1\delta + h_2 + h_3\big(\delta + a_3\delta + \hat{p}_1(\delta) + \delta\big) \\ &= (h_1 + 2h_3 + h_3a_3)\delta + h_3\hat{p}_1(\delta) + h_2 \\ &=: h_5\delta + h_4(\delta) + h_2 \end{aligned} \tag{81}$$

where

$$\begin{aligned} h_1 &:= \sup_{\|z_1\|\leq\delta}\|D^{-1}C\| < \infty, \\ h_2 &:= \sup_{\|Z_1\|\leq\delta}\|D^{-1}G\| < \infty, \\ h_3 &:= \sup_{\|Z_1\|\leq\delta}\|D^{-1}K\| < \infty, \end{aligned} \tag{82}$$

$$h_4(\delta) := h_3\hat{p}_1(\delta) < \infty,$$
$$h_5 := h_1 + 2h_3 + h_3a_3.$$

From (79) we have

$$\begin{aligned}\|\dot{p}_1\| &\le \bar{\eta}_1\delta + \bar{\eta}_2(h_5\delta + h_4(\delta) + h_2)\\ &=: \bar{\eta}_3\delta + \bar{\eta}_2h_5(\delta) + \bar{\eta}_2h_2,\end{aligned} \tag{83}$$

where

$$\bar{\eta}_3 := \bar{\eta}_1 + \bar{\eta}_2h_5. \tag{84}$$

Therefore, by the results above

$$\begin{aligned}\|\dot{u}_1\| &\le \|\bar{K}^{-1}\|\|K_{p1} + \beta_1S_1\|\|\dot{Z}_1\|\\ &\quad + \|\bar{K}^{-1}\|\|K_{v1} + \beta_1I\|\|\ddot{Z}_1\| + \|\bar{K}^{-1}\|\|\dot{p}_1\|\\ &\le h_6\delta + h_7(\delta) + h_8,\end{aligned} \tag{85}$$

where

$$\begin{aligned}h_6 &:= \|\bar{K}^{-1}\|(\|K_{p1} + \beta_1S_1\| + h_5\|K_{v1} + \beta_1I\| + \bar{\eta}_3),\\ h_7(\delta) &:= \|\bar{K}^{-1}\|(\|K_{p1} + \beta_1I\| + \bar{\eta}_2)h_4(\delta),\\ h_8 &:= \|\bar{K}^{-1}\|h_2(\|K_{v1} + \beta_1I\| + \bar{\eta}_2).\end{aligned} \tag{86}$$

The $\|x_2\|$ is bounded:

$$\begin{aligned}\|x_2\| &\le \|z_2\| + \|u_1\| + \|\dot{u}_1\|\\ &\le \delta + \|u_1\| + \|\dot{u}_1\|\\ &\le \delta + a_3\delta + \hat{p}_1(\delta) + h_6\delta + h_7(\delta) + h_8\\ &= H_1\delta + H_2(\delta) + H_3,\end{aligned} \tag{87}$$

where

$$\begin{aligned}H_1 &:= 1 + a_3 + h_6,\\ H_2(\delta) &:= \hat{p}_1(\delta) + h_7(\delta),\\ H_3 &:= h_8.\end{aligned} \tag{88}$$

Therefore, we get

$$\begin{aligned}\|x\| &= (\|x_1\|^2 + \|x_2\|^2)^{\frac{1}{2}}\\ &\le \left(\delta^2 + (H_1\delta + H_2(\delta) + H_3)^2\right)^{\frac{1}{2}}\\ &=: w(\delta) < \infty.\end{aligned} \tag{89}$$

We investigate the uniform boundedness of x based on the performance of z. As δ is replaced by $d_z(r_z)$, (89) leads to

$$\|x\| \le w(d_z(r_z)). \tag{90}$$

Given any $r_z > 0$, with $\|z_0\| \leq r_z$, there is a $r_x = d_x^{-1}(w(\delta))$ with $\|x_0\| \leq r_x$. Let

$$d_x(r_x) = \begin{cases} R_x\sqrt{\frac{\gamma_2}{\gamma_1}} & \text{if } r_x \leq R_x \\ r_x\sqrt{\frac{\gamma_2}{\gamma_1}} & \text{if } r_x > R_x, \end{cases} \tag{91}$$

where $R_x = w(R_z)$. By (91), $\|z(t)\| \leq d_z(r_z)$ implies $\|x(t)\| \leq d_x(r_x)$. Let $\underline{d}_x = R_x\sqrt{\gamma_2/\gamma_1}$. The uniform ultimate boundedness of $x(t)$ also follows by choosing, for $\bar{d}_x > \underline{d}_x$,

$$T_x(\bar{d}_x, r_x) = \begin{cases} 0 & \text{if } r_x \leq \bar{d}_x\sqrt{\frac{\gamma_2}{\gamma_1}} \\ \frac{\gamma_2 r_x^2 - \gamma_1^2\gamma_2^{-1}\bar{d}_x^2}{\gamma_1\gamma_2^{-1}\gamma_3\bar{d}_x^2 - \bar{k}} & \text{otherwise.} \end{cases} \tag{92}$$

For the uniform stability, let $\delta_x(\bar{d}_x) = R_x$; then $\|x_0\| \leq \delta(\bar{d}_x)$ implies $\|x(t)\| \leq \bar{d}_x$.

Theorem 2 *Subject to Assumptions 1–3, the control u given by (28) renders the original dynamic system N_1 and N_2 defined by (7) practically stable.*

Proof The practical stability has been shown as above. Q.E.D.

Lemma 1 If either the gravitational force $G(q_1)$ is absent (i.e., $G(q_1) \equiv 0$) or the coordinate q_1 is selected in such a way $G(q_1) \to 0$ as $q_1 \to 0$, then the original system is practically stable. Furthermore, the size of the uniform ultimate boundedness region can be made arbitrarily small by a suitable choice of ϵ_1 and ϵ_2.

Proof The practical stability has been shown above. The constant $\underline{d}_x$, which determines the size of the uniform ultimate boundedness ball and stability ball, approaches 0 as R_x approaches 0. This is so since as $\epsilon_1, \epsilon_2 \to 0$, $\bar{k} \to 0$ and hence $\underline{d}_z \to 0$, so does $\delta \to 0$. This yields $h_2 \to 0$, hence $H_2(\delta) \to 0$ and $H_3 \to 0$. Therefore $\underline{d}_x = w(\underline{d}_z) \to 0$. Q.E.D.

Remark Similar results as in Theorem 1 and 2 and Lemma 1 hold for the time-varying uncertainty case.

8 Conclusions

A robust control scheme is proposed for flexible joint manipulators. The features of the manipulators, which distinguish the current problem from others, are that the dynamic system is nonlinear, uncertain, and mismatched. The suggested control scheme is designed via a two-step procedure. In essence, the mismatched system is first transformed to a matched one. The resulting control is for the transformed system. The performance requirement is, however, specified for the original system. The analysis of the original system performance based on that of the transformed system is demonstrated.

Acknowledgement

The authors are grateful for valuable comments made by Prof. G. Leitmann (UC Berkeley).

References

[1] Chen, Y.H. and Leitmann, G. (1989) Robustness of uncertain systems in the absence of matching assumptions, *International Journal of Control*, **45**, 1527–1542.

[2] Corless, M. (1993) Control of uncertain nonlinear systems. *Journal of Dynamic Systems, Measurement, and Control*, **115**, 362–372.

[3] Corless, M.J. and Leitmann, G. (1981) Continuous state feedback guaranteeing uniform ultimate boundedness for uncertain dynamic systems. *IEEE Transactions on Automatic Control*, **26**(5), 1139–1143.

[4] Craig, J.J. (1989) *Introduction of Robotics: Mechanics and Control*, Second Edition. Addison-Wesley.

[5] Freeman, R.A. and Kokotovic, P.V. (1992) Design and comparison of globally stabilizing controllers for an uncertain nonlinear system, in *Systems, Models, and Feedback: Theory and Applications*, A. Isidori, T.J. Tarn (eds.). Birkhauser, Boston.

[6] Han, M.C. and Chen, Y.H. (1993) Robust control design for flexible-joint manipulators: singular perturbation approach. *Proceedings of the 32nd IEEE Conference on Decision and Control*, San Antonio, TX.

[7] Leitmann, G. (1993) On one approach to the control of uncertain systems. *Journal of Dynamic Systems, Measurement, and Control*, **115**, 373–380.

[8] Spong, M.W. (1990) The control of flexible joint robots: A survey, In *New Trends and Applications of Distributed Parameter Control system*, Lecture Notes in Pure and Applied Mathematics, G. Chen, E.B. Lee, W. Littman, and L. Markus. Eds. Marcel Dekker Publishers, NY.

[9] Spong, M.W. and Vidyasagar, M. (1989) *Robot Dynamics and Control*. Wiley, NY.

4 CONTROL UNDER LACK OF INFORMATION

ANDREW N. KRASOVSKII

Department of Mathematics and Mechanics, Ural State University, Ekaterinburg 620083, Russia

This paper is devoted to feedback control problems for a dynamical system in a situation of uncertain information about the input disturbances, or in a situation of conflict. Problems of finding controls minmax and maxmin with respect to a given quality index are formalized into a zero-sum differential game of two players. Our considerations are based on an approach developed in the Ural State University, Ekaterinburg (former Sverdlovsk). We describe a structure of optimal control strategies and characterize the optimal ensured result, or, equivalently, the value of the game. The quality indexes estimate motions and control actions over the full period of control. We give a new classification of differential games according to a type of a quality index and formulate existence theorems for the values and saddle points — pairs of optimal strategies.

1 Motion Equation

Assume that a motion of a control system is described by a differential equation

$$\dot{x} = A(t)x + f(t, u, v), \quad t_0 \leq t \leq \theta, \quad u \in P, \quad v \in Q \tag{1}$$

where x is an n-dimensional phase vector, t is time, t_0 and θ are fixed instants, u is an r-dimensional vector of control, and v is an s-dimensional vector of disturbance. The matrix-valued function $A(t)$ and the vector function $f(t, u, v)$ are piecewise-continuous in t, and $f(t, u, v)$, for every fixed t, is continuous in u and v varying in sets P and Q, respectively. In (1) P and Q are compact describing the abilities of a regulator U generating controls u (or actions of the first player), and a regulator V generating disturbances v (or actions of the second player).

We assume that function $f(\cdot)$ satisfies the saddle point condition in the "small game", i.e.

$$\min_{u\in P}\max_{v\in Q}\langle l, f(t,u,v)\rangle = \max_{v\in Q}\min_{u\in P}\langle l, f(t,u,v)\rangle \tag{2}$$

for every n-dimensional vector l.

The motion $\{x[t], t_* \le t \le t^*\}$ of system (1) on a time interval $[t_*, t^*] \subset [t_0, \theta]$ generated by a (Borel measurable) control $\{u[t] \in P, t_* \le t \le t^*\}$ and a (Borel measurable) disturbance $\{v[t] \in Q, t_* \le t \le t^*\}$ from an initial position $\{t_*, x_*\}$ is defined to be the (Caratheodory) solution of the differential equation

$$\dot{x}[t] = A(t)x[t] + f(t, u[t], v[t]), \quad t_* \le t \le t^*, \quad x[t_*] = x_* \tag{3}$$

We shall deal with admissible positions $\{t, x[t]\}$ satisfying the inclusion $\{t, x[t]\} \in G$ where G is some compact set in the product of the time interval $[t_0, \theta]$ and n-dimensional space. Functions $x[t]$ (3) are, obviously, Lipschitz in t.

2 Positional Quality Index

We consider minmax and maxmin (with respect to u and v) problems for an index evaluating the quality of control processes. In a general case a quality index γ is defined as

$$\gamma = \gamma(\{x[t]; u[t]; v[t], t_* \le t \le \theta\}) \tag{4}$$

thus depending on motion, $x[t]$, and realizations of control, $u[t]$, and disturbance, $v[t]$. In (4) $t_* \in [t_0, \theta]$ is the initial time for a control process and θ is the fixed terminal time given in (1).

In this paper we consider four groups of indexes γ (4).

At first, we restrict ourselves to the case where the quality index depends on a motion $\{x[t], t_* \le t \le \theta\}$ of the controlled plant (1) only, i.e., $\gamma = \gamma(\{x[t], t_* \le t \le \theta\})$. We shall consider problems of designing controls u aimed at minimizing γ, and disturbances v aimed at maximizing γ. We employ the feedback control pattern in which at every time $t \in [t_*, \theta]$ a control u (respectively, a disturbance v) is worked out using the current state of object (1), $x[t]$. In a slightly modified setting, control u (respectively, disturbance v) is updated at discrete instants t_i, $i = 1, \ldots, k+1$, that form a finite partition of the time interval $[t_*, \theta]$ ($t_* = t_1, \theta = t_{k+1}$); at time t_i position $\{t_i, x[t_i]\}$ is utilized. The problems can be formalized within the framework of theory of positional differential games [1].

Now we consider a specific form of quality indexes. We call such quality indexes positional. Namely, a quality index γ is called positional if it can be represented in the form

$$\begin{aligned}\gamma(\{x[t], t_* \le t \le \theta\}) &= \beta(\{x[t], t_* \le t \le t^*\}, \alpha),\\ \alpha &= \gamma(\{x[t], t^* \le t \le \theta\})\end{aligned} \tag{5}$$

($t^* \in (t_*, \theta]$) where the functional $\beta(\cdot)$ is continuous and nondecreasing in α. In particular, the following functionals are positional:

$$\gamma = \sigma(x[\theta]),$$

$$\gamma = \int_{t_*}^{\theta} \psi(t, x[t])dt + \sigma(x[\theta]),$$

$$\gamma = \sum_{i=g}^{N} |x[t^{[i]}]|,$$

$$\gamma = \max_{t_* \leq t \leq \theta} |x[t]|,$$

$$\gamma = \sum_{i=g}^{N} |x[t^{[i]}] - c^{[i]}|,$$

$$\gamma = \max_{t_* \leq t \leq \theta} |x[t] - c[t]|$$

where ψ and σ are given continuous functions Lipschitz in x, symbol $|x|$ denotes the Euclidean norm of x, $t^{[i]}$ are given time instants, $c^{[i]}$ are fixed vectors, and $c[t]$ is a continuous vector function. We attribute the quality indexes of the form (5) to the first group of functionals γ (4).

3 Pure Positional Strategies

A pure positional strategy $u(\cdot)$ of the first player (respectively, a pure positional strategy $v(\cdot)$ of the second player) is understood as a rule which, for every admissible position $\{t, x\}$ of object (1), assigns a control action u (respectively, a disturbance action v) provided an auxiliary accuracy parameter ϵ is chosen.

More accurately, we call a function

$$u(\cdot) = \{u(t, x, \epsilon_u) \in P,\ t \in [t_0, \theta],\ \epsilon_u > 0\} \tag{6}$$

a pure positional strategy of the first player. Here $\epsilon_u > 0$ is an accuracy parameter to be chosen by the first player (see [1], [2]).

A feedback control law U is determined by three components: a strategy $u(\cdot)$ (6), an accuracy parameter $\epsilon_u > 0$, and a partition $\Delta = \{t_* = t_1, \ldots, t_{k+1} = \theta\}$. The motion

$$\{x_{U,v}[t], t_* \leq t \leq \theta, x_{U,v}[t_*] = x_*\} \tag{7}$$

generated by the control law U and a disturbance $\{v[t] \in Q, t_* \leq t \leq \theta\}$ from an initial position $\{t_*, x_*\}$ is determined as the step-by-step solution of the differential equation

$$\dot{x}_{U,v}[t] = A(t)x_{U,v}[t] + f(t, u(t, x_{U,v}[t_i], \epsilon_u), v[t]),\ t_i \leq t \leq t_{i+1}, \quad i = 1, \ldots, k. \tag{8}$$

Similarly, a pure positional strategy of the second player is understood as a function

$$v(\cdot) = \{v(t, x, \epsilon_v) \in Q,\ t \in [t_0, \theta]\ \epsilon_v > 0\},$$

and a feedback control law V of the second player is determined by a strategy $v(\cdot)$, an accuracy parameter $\epsilon_v > 0$, and a partition $\Delta = \{t_* = t_1, \ldots, t_{k+1} = \theta\}$. The motion

$$\{x_{V,u}[t], t_* \leq t \leq \theta, x_{V,u}[t_*] = x_*\}$$

generated by the control law V and a control of the first player $\{u[t] \in P, t_* \leq t \leq \theta\}$ from an initial position $\{t_*, x_*\}$ is determined as the step-by-step solution of the differential equation

$$\dot{x}_{U,v}[t] = A(t)x_{V,u}[t] + f(t, u[t], v(t, x_{V,u}[t], \epsilon_v)),\ t_i \leq t \leq t_{i+1},\quad i = 1, \ldots, k.$$

4 Problem Statement for Positional Quality Index

Consider a positional quality index γ (5). Each collection $\{\{t_*, x_*\}, U, \{v[t] \in Q, t_* \leq t \leq \theta\}$ determines, via motion (8), a unique value $\gamma(\{x_{U,v}[t], t_* \leq t \leq \theta\})$. The ensured result for a control law U of the first player at an initial position $\{t_*, x_*\}$ (with respect to the quality index γ) is defined as follows:

$$\rho(t_*, x_*, U) = \sup_{\{v[t] \in Q, t_* \leq t \leq \theta\}} \gamma(\{x_{U,v}[t], t_* \leq t \leq \theta\}). \tag{9}$$

The ensured result for strategy $u(\cdot)$ (6) of the first player at $\{t_*, x_*\}$ is given by

$$\rho(t_*, x_*, u(\cdot)) = \lim_{\epsilon \to 0} \lim_{\delta \to 0} \sup_{\Delta} = \rho(t_*, x_*, U). \tag{10}$$

According to (10), for arbitrary $\zeta > 0$ there are $\epsilon(\zeta) > 0$ and $\delta(\zeta, \epsilon) > 0$ such that the unequality

$$\gamma(\{x_{U,v}[t], t_* \leq t \leq \theta\}) \leq \rho(t_*, x_*, u(\circ)) + \zeta$$

holds for motion (8) whenever $\epsilon_u \leq \epsilon(\zeta)$ and $t_{i+1} - t_i \leq \delta(\zeta, \epsilon)$. We call a strategy $u^0(\cdot)$ optimal if the equality

$$\rho(t_*, x_*, u^0(\circ)) = \min_{u(\cdot)} \rho(t_*, x_*, u(\circ)) = \rho_u^0(t_*, x_*)$$

holds for every initial position $\{t_*, x_*\}$; value $\rho_u^0(t_*, x_*)$ is called then the optimal ensured result for the first player at $\{t_*, x_*\}$.

For the second player symmetric definitions are assumed. The ensured result for strategy $v(\cdot)$ of the second player at an initial position $\{t_*, x_*\}$ is given by

$$\rho(t_*, x_*, u(\cdot)) = \lim_{\epsilon \to 0} \lim_{\delta \to 0} \sup_{\Delta} \rho(t_*, x_*, V);$$

here

$$\rho(t_*, x_*, V) = \sup_{\{u[t] \in P,\, t_* \leq t \leq \theta\}} \gamma(\{x_{V,u}[t], t_* \leq t \leq \theta\})$$

is the ensured result at $\{t_*, x_*\}$ for a control law V incorporating strategy $v(\cdot)$. A strategy $v^0(\cdot)$ of the second player is optimal if

$$\rho(t_*, x_*, v^0(\cdot)) = \max_{v(\cdot)} \rho(t_*, x_*, v(\cdot)) = \rho_v^0(t_*, x_*)$$

for every $\{t_*, x_*\}$; $\rho_v^0(t_*, x_*)$ is called the optimal ensured result for the second player at $\{t_*, x_*\}$.

The guarantee optimal control problems of the first and second players consist in finding optimal strategies $u^0(\cdot)$ and $v^0(\cdot)$, respectively. A pair of these problems is treated as a zero-sum differential game of two persons, [1], [2], [3]. Using the terminology of game theory, we say that a pair of optimal strategies, $\{u^0(\cdot), v^0(\cdot)\}$, forms a saddle point in the game and provides a value function $\rho^0(t_*, x_*)$ if the equality

$$\rho_u^0(t_*, x_*) = \rho_v^0(t_*, x_*) = \rho^0(t_*, x_*) \tag{11}$$

holds for every position $\{t_*, x_*\}$. The saddle point $\{u^0(\cdot), v^0(\cdot)\}$ is uniform in the sense that strategies $u^0(\cdot)$ and $v^0(\cdot)$ do not depend on an initial position.

We are interested in stating the existence of a value function $\rho^0(t_*, x_*)$ and specifying the structure of a uniform saddle point $\{u^0(\cdot), v^0(\cdot)\}$. Relevant results are presented below.

5 Extremal Shift Method

In this section so-called extremal strategies $u^e(\cdot) = u^e(t, x, \epsilon_u)$ and $v^e(\cdot) = v^e(t, x, \epsilon_v)$ are introduced.

Suppose that we know the value function $\rho^o(t, x)$, or, at least, are able to compute $\rho^o(t, x)$ for each $\{t, x\}$. Let us describe the extremal strategy $u^e(\cdot)$ of the first player. Take a position $\{t, x\}$ and choose a small $\epsilon_u > 0$. An associated control action $u^e(t, x, \epsilon_u)$ is defined as follows. Denote by $K^u(\epsilon_u, t, x)$ the set of all positions $\{t, w\}$ such that

$$|x - w| \leq (\epsilon_u + \epsilon_u(t - t_0))^{1/2} \exp\{\lambda(t - t_0)\}.$$

Here $\lambda > 0$ is an appropriate constant (see [1] for details). As ϵ_u is small, the w components of positions $\{t, w\} \in K^u(\epsilon_u, t, x)$ are located in a small Euclidean neighborhood of x. Select a point $w^u(t, x, \epsilon_u) \in K^u(\epsilon_u, t, x)$ such that

$$\rho^0(t, w^u(t, x, \epsilon_u) = \min_{\{t,w\} \in K} u_{(\epsilon_u, t, x)} \rho^0(t, w). \tag{12}$$

We call $w^u(t,x,\epsilon_u)$ an accompanying point. The control action $u^e = u^e(t,x,\epsilon_u) \in P$ is determined by

$$\max_{v\in Q}\langle(x - w^u(t,x,\epsilon_u)), f(t,u^e,v)\rangle = \min_{u\in P}\max_{v\in Q}\langle(x - w^u(t,x,\epsilon_u), f(t,u,v)\rangle. \tag{13}$$

We call (13) the condition of the extremal shift to the accompanying point for the first player.

For the second player an accompanying point $w^v(t,x,\epsilon_v)$ is defined by

$$\rho^0(t, w^v(t,x,\epsilon_v)) = \max_{w\epsilon K} v_{(\epsilon_v,t,x)}\rho^0(t,w) \tag{14}$$

and the extremal strategy $v^e(\cdot)$ is determined by the following condition of the extremal shift:

$$\max_{u\in P}\langle(x - w^v(t,x,\epsilon_v)),\ f(t,u,v^e)\rangle = \min_{v\in Q}\max_{u\in P}\langle(x - w^v(t,x,\epsilon_v)),\ f(t,u,v)\rangle. \tag{15}$$

6 Existence of Solution for Positional Quality Index

The following result is true.

Theorem 1 *The zero-sum differential game for system (1), (2) with the positional quality index $\gamma(\{x[t], t_* \le t \le \theta\})$ (5) has a value $\rho^0(t,x)$, and the pair of extremal strategies, $\{u^e(\cdot), v^e(\cdot)\}$, is a uniform saddle point in the game.*

A proof of Theorem 1 follows a slightly modified argument pattern from [1].

As we have seen, optimal strategies $u^0(\cdot) = u^e(\cdot)$ and $v^0(\cdot) = v^e(\cdot)$ can be efficiently constructed, given the value function of the game, $\rho^0(t,x)$. Some methods for computing $\rho^0(t,x)$ (or $\rho^0(t,w)$, as long as procedures (12)–(15) are concerned) have been reviewed in [1]. We shall not repeat here any computational schemes but rather formulate the following general statement.

Theorem 2 *Let function $\rho(t,w)$ satisfy the next conditions:*

1) $\rho(t,w)$ is Lipshitz in w, i.e.,

$$|\rho(t,w^{(1)}) - \rho(t,w^{(2)})| \le L|w^{(2)} - w^{(1)}|, \quad L = const, \quad t, w^{(i)} \in G,$$

2) $\gamma(\{w[t], t = \theta\}) = \rho(\theta, w[\theta])$ holds for $\{\theta, w[\theta]\} \in G$,

3) $\rho(t,w)$ is u-stable (see [1], p. 61), i.e., for every position $\{\tau_*, w_*\} \in G$, every $\tau^* > \tau_*$, every admissible control $\{v[t] \in Q, \tau_* \le t \le \tau^*\}$, of the second player, and every $\epsilon > 0$ there is an admissible control $\{u[t] \in P, \tau_* \le t \le \tau^*\}$ of the first player such that for the motion $\{x[t], \tau_* \le t \le \tau^*\}$ generated by controls $\{u[t] \in P,$

$\tau_* \leq t \leq \tau^*\}$ and $\{v[t] \in Q, \tau_* \leq t \leq \tau^*\}$ from the initial position $\{\tau_*, w_*\}$ it holds that

$$\beta(\{x[t], \tau_* \leq t \leq \tau^*\}, \rho(\tau^*, x[\tau^*]) + 2\epsilon) \leq \rho(\tau_*, w_*) + 2\epsilon$$

where$\beta(\cdot)$is the functional from (5),

4) *$\rho(t, w)$ is v-stable (see [1], p. 61), i.e., for every position $\{\tau_*, w_*\} \in G$, every $\tau^* > \tau_*$, every admissible control $\{u[t] \in P, \tau_* \leq t \leq \tau^*\}$, of the first player, and every $\epsilon > 0$ there is an admissible control $\{v[t] \in Q, \tau_* \leq t \leq \tau^*\}$ of the second player such that for the motion $\{x[t], \tau_* \leq t \leq \tau^*\}$ generated by controls $\{u[t] \in P, \tau_* \leq t \leq \tau^*\}$ and $\{v[t] \in Q, \tau_* \leq t \leq \tau^*\}$ from the initial position $\{\tau_*, w_*\}$ it holds that*

$$\beta(\{x[t], \tau_* \leq t \leq \tau^*\}, \rho(\tau^*, x[\tau^*]) - 2\epsilon) \geq \rho(\tau_*, w_*) - 2\epsilon.$$

Then $\rho^0(t, x) = \rho(t, x)$ is the value in the differential game for the system (1), (2) with the positional quality index γ (5).

This fact has been established in [2] where also a method to construct $\rho(t, w)$ via so-called Q-procedures was suggested.

7 Differential Game with Quasipositional Quality Index

Now let the quality index γ (4) be not necessarily positional, i.e., the representation (5) may not take place. Consider, for example, the following variants:

$$\gamma_I = \gamma_1 + J, \quad \gamma_{II} = \gamma_1 + \gamma_2, \quad \gamma_{III} = \gamma_2 + J, \quad \gamma_{IV} = \gamma_1 + \gamma_2 + J$$

where

$$\gamma_1 = \sum_{i=g}^{N} |x[t^{[i]}]|, \quad \gamma_2 = \max_{i=1,\ldots,N} |x[t^{[i]}]|, \quad J = \int_{t_*}^{\theta} \phi(t, u[t], v[t]) dt.$$

Here the function $\phi(t, u, v)$ is piecewise-continuous in t and continuous within u and v. We call these functionals quasi-positional.

We assume that the following modified saddle point condition in the "small game" is satisfied: for arbitrary n-dimensional vector l and arbitrary scalar q it holds that

$$\min_{u \in P} \max_{v \in Q} [\langle l, f(t, u, v) \rangle + q\phi(t, u, v)] = \max_{v \in Q} \min_{u \in P} [\langle l, f(t, u, v) \rangle + q\phi(t, u, v)]. \tag{18}$$

The expressions for γ_j, $j = I, II, III, IV$ determine four groups of quasipositional quality indexes. It has been proved in [1] that for each of these groups there is a specific sufficient information image for players' strategies $u(\cdot)$ and $v(\cdot)$. A sufficient information image $y_{(j)}[t]$ associated with γ_j is a function of the current history of a control process such that the differential game with the quality index γ_j has a saddle point $\{u^0_{(j)}(\cdot), v^0_{(j)}(\cdot)\}$ in players' strategies of the form $u_{(j)}(y_{(j)}[t], \epsilon_u), v_{(j)}(y_{(j)}[t], \epsilon_v)$. For a previously considered positional quality index γ (5) a sufficient information image is $y[t] = \{t, x[t]\}$, since, as we have seen, a saddle point can be constructed in the form $\{u^0(t, x[t], \epsilon_u), v^0(t, x[t], \epsilon_v)\}$.

Thus, we classify quality indexes into four groups in accordance with the associated sufficient information images. The first group comprises all quality indexes γ for which a sufficient information image $y_{(I)}[t]$ is the current position $\{t, x[t]\}$, i.e., $y_{(I)}[t] = \{t, x[t]\}$. The second group comprises all γ for which a sufficient information image $y_{(II)}[t]$ is the history of the motion up to the current time t, i.e., $y_{(II)}[t] = \{x[\tau], t_* \leq \tau \leq t\}$. The third group consists of the quality indexes γ for which a sufficient information image is the extended position, $y_{(III)} = \{t, x[t], x_{n+1}[t]\}$, where

$$x_{n+1}[t] = \int_{t_*}^{t} \phi(\tau,\ u[\tau],\ v[\tau])\ d\tau$$

Finally, the fourth group is composed of all γ for which a sufficient information image is the history and extended position, $y_{(IV)}[t] = \{\{x[\tau], t_* \leq \tau \leq t\}, x_{n+1}[t]\}$.

For the quality indexes γ_j, $j = I,\ II,\ III,\ IV$ the following theorem is true.

Theorem 3 *The zero-sum differential game of two persons for system (1), (16) with the quality index* γ_j *has a value* $\rho^0_{(j)}(y_{(j)}[t])$ *and a saddle point* $\{u^0_{(j)}(y_{(j)}[t], \epsilon_u), v^0_{(j)}(y_{(j)}[t], \epsilon_v)\}$.

We omit the proof which can be performed using modified arguments of [1]. We only note that optimal strategies can again be constructed as extremal ones, and, for each quality index γ_j, accompanying points for the extremal strategies are sought in the spaces of extended variables associated with the relevant sufficient information images. For the practical calculation of the value functions, the method of upper convex hulls, [3], can be implemented.

References

[1] Krasovskii, A.N. and Krasovskii, N.N. (1995) *Control Under Lack of Information*. Birkhauser, Boston, USA, 320p.

[2] Krasovskii, A.N. (1980) On positional minimax control, *J. Appl. Math. Mech.*, **44**(4), 602–610 (in Russian).

[3] Krasovskii, A.N. (1987) Construction of mixed strategies based on stochastic programs, *J. Appl. Math. Mech.*, **51**(2), 186–192 (in Russian).

5 A DIFFERENTIAL GAME WITH HEREDITARY INFORMATION*

NIKOLAI YU. LUKOYANOV

Department of Dynamic Systems, Institute of Mathematics and Mechanics, Ural Branch of Russian Academy of Sciences, Ekaterinburg 620219, Russia

For a dynamical system with uncertain input disturbances a problem of guarantee optimal closed-loop control is considered. A quality index is determined by a functional on a system's motions. The problem is formalized into a two-person zero-sum differential game, [1], [2]. We focus on a situation where the controls base essentially on the history of a motion. The problem is resolved via the construction of upper convex hulls for auxiliary functions in a multidimensional space (see [1], [3]). A method of reducing multidimensional constructions to operations in smaller dimensions is described.

The problem statement and solution approach (adjoining [1]) have been proposed by N.N. Krasovskii.

1 Statement of the Problem

Let us consider a system

$$dx/dt = A(t)x + f(t, u, v), \quad 0 \leq t_*^0 \leq t \leq \vartheta. \tag{1.1}$$

Here x is an n-dimensional state vector, u is an r-dimensional control vector, and v is an s-dimensional disturbance vector; t_*^0 and ϑ are fixed instants; $A(t)$ is a continuous matrix-valued function and $f(t, u, v)$ is a continuous vector-valued function; values u and v are subject to

$$u \in P, \quad v \in Q. \tag{1.2}$$

* The research described in this publication has been supported in part from Grant NMS300 of the International Science Foundation.

Here P and Q are compacta. We assume that the saddle point condition in the "small game" (see [1], [2]) is satisfied, i.e., for any n-dimensional vector m and time $t \in [t_*^0, \vartheta]$ one has

$$\min_{u\in P}\max_{v\in Q}\langle m, f(t,u,v)\rangle = \max_{v\in Q}\min_{u\in P}\langle m, f(t,u,v)\rangle. \tag{1.3}$$

Here and in what follows symbol $\langle\cdot,\cdot\rangle$ denotes the scalar product.

Admissible control and disturbance realizations are Borel measurable functions $u[t_*^0[\cdot]\vartheta) = \{u[t] \in P, t_*^0 \le t < \vartheta\}$ and $v[t_*^0[\cdot]\vartheta) = \{v[t] \in Q,\ t_*^0 \le t < \vartheta\}$. Given a starting state $x[t_*^0]$, realizations $u[t_*^0[\cdot]\vartheta)$ and $v[t_*^0[\cdot]\vartheta)$ generate a unique absolutely continuous motion $x[t_*^0[\cdot]\vartheta] = \{x[t],\ t_*^0 \le t \le \vartheta\}$ being a solution of the equation (1.1) with $u = u[t]$, $v = v[t]$.

Let us introduce a quality index of a control process, $\gamma = \gamma(x[t_*^0[\cdot]\vartheta])$.

Take a natural N, instants $t^{[i]} \in [t_*^0, \vartheta]$, $t^{[i+1]} > t^{[i]}$, $i = 1, \ldots, N-1$, $t^{[N]} = \vartheta$, and constant $(p^{[i]} \times n)$-matrices $D^{[i]}$, $1 \le p^{[i]} \le n, i = 1, \ldots, N$. The collection $\{D^{[1]}x[t^{[1]}], \ldots, D^{[N]}x[t^{[N]}]\}$ forms a p-dimensional vector, $p = \sum_{i=1}^{N} p^{[i]}$. Let $\mu(\cdot)$ be a norm in R^p. We set

$$\gamma = \gamma\big(x[t_*^0[\cdot]\vartheta]\big) = \mu\Big(\Big\{D^{[1]}x[t^{[1]}], \ldots, D^{[N]}x[t^{[N]}]\Big\}\Big). \tag{1.4}$$

The quality index γ (1.4) can either be given originally, or introduced as an approximation to a $\gamma_*(x[t_*^0[\cdot]\vartheta])$ estimating a continuum of values of $x[t]$.

We consider two problems. The first is to find a closed-loop control u minimizing γ, and the second is to find a closed-loop disturbance v maximizing γ.

In accordance with [1], these problems determine a two-person zero-sum differential game in which u is the action of the first player and v is that of the second player. This game has a value $\rho^o(x[t_*^0[\cdot]t_*])$ and a saddle point $\{u^o(x[t_*^0[\cdot]t], \epsilon), v^o(x[t_*^0[\cdot]t], \epsilon)\}$. Here $x[t_*^0[\cdot]t_*] = \{x[\tau], t_*^0 \le \tau \le t_*\}$ $(t_*^0 \le t_* < \vartheta)$ is the initial history of a motion, $x[t_*^0[\cdot]t] = \{x[\tau], t_*^0 \le \tau \le t\}$ is the history up to a current time t, and ϵ is an accuracy parameter, [1], [2]. For every initial history $x[t_*^0[\cdot]t_*]$ a closed-loop control law U forming a control realization step-by-step by

$$u[t] = u^o(x[t_*^0[\cdot]t_j^{(u)}], \epsilon_u),\ t_j^{(u)} \le t < t_{j+1}^{(u)},\ j = 1, \ldots, k_u,\ t_1^{(u)} = t_*,\ t_{k_u}^{(u)} = \vartheta$$

ensures

$$\gamma \le \rho^o(x[t_*^0[\cdot]t_*]) + \zeta$$

for arbitrary $\zeta > 0$; provided parameter $\epsilon_u > 0$ and step $\delta_u = \max_j(t_{j+1}^{(u)} - t_j^{(u)})$ are sufficiently small. On the other hand, a closed-loop disturbance law V forming a disturbance realization step-by-step by

$$v[t] = v^o(x[t_*^0[\cdot]t_j^{(v)}], \epsilon_v),\ t_j^{(v)} \le t < t_{j+1}^{(v)},\ j = 1, \ldots, k_v,\ t_1^{(v)} = t_*,\ t_{k_v}^{(v)} = \vartheta$$

ensures

$$\gamma \ge \rho^o(x[t_*^0[\cdot]t_*]) - \zeta$$

provided $\epsilon_v > 0$ and $\delta_v = \max_j(t_{j+1}^{(v)} - t_j^{(v)})$ are sufficiently small.

Optimal strategies $u^o(x[t_*^0[\cdot]t],\epsilon)$ and $v^o(x[t_*^0[\cdot]t],\epsilon)$ which constitute a saddle point can be constructed as extremal ones with respect to the functional $\rho^o(x[t_*^0[\cdot]t])$ (see [1]).

Remark 1 In many typical situations optimal strategies utilize partial information on current histories. For example, if γ (1.4) is a positional functional, [1], then only the current positions, $\{t, x[t]\}$, can be taken into account.

Remark 2 Condition (1.3) ensures that the differential game under consideration has a saddle point in pure strategies. If (1.3) is violated, a saddle point is transferred to the class of mixed strategies, [1]. However, auxiliary constructions described in this paper do not change in this situation.

Thus, in order to form an optimal control strategy $u^o(x[t_*^0[\cdot]t_j^{(u)}],\epsilon_u)$ (or disturbance strategy $v^o(x[t_*^0[\cdot]t_j^{(v)}],\epsilon_v)$) that guarantees the optimal result $\rho^o(x[t_*^0[\cdot]t_*])$, it is sufficient to compute $\rho^o(x[t_*^0[\cdot]t])$, the value of the game (1.1)–(1.4), for every admissible history $x[t_*^0[\cdot]t]$.

Our objective is to provide an effective procedure for computing these values.

First, we give a functional interpretation for the control process (1.1)–(1.4).

2 Functional Interpretation

A functional position corresponding to a history $x[t_*^0[\cdot]t]$ $(t\in[t_*^0,\vartheta])$ is identified with a collection $\{t,\widehat{\mathbf{z}}[t]\}$, where

$$\widehat{\mathbf{z}}[t]=(x[t],\widehat{\mathbf{x}}[t]),\quad \widehat{\mathbf{x}}[t]=\{\widehat{x}^{[1]}[t],\ldots,\widehat{x}^{[N]}[t]\},\tag{2.1}$$

$$\widehat{x}^{[i]}[t]=\begin{cases}D^{[i]}x[t^{[i]}], & t^{[i]}\le t\\ D^{[i]}X(t^{[i]},t)x[t], & t<t^{[i]}\end{cases}.$$

Here $X(\tau,t)$ is the fundamental solution matrix for the equation $dx/d\tau=A(\tau)x$. Now the quality index γ (1.4) can be rewritten in the form $\gamma=\mu(\widehat{\mathbf{x}}[\vartheta])$.

The evolution of a functional position $\{t,\widehat{\mathbf{z}}[t]=(x[t],\widehat{\mathbf{x}}[t])\}$ is described by equations (1.1) and

$$d\widehat{\mathbf{x}}[t]/dt=\widehat{\mathbf{f}}(t,u,v),\quad t_*^0\le t\le\vartheta\tag{2.2}$$

where

$$\widehat{\mathbf{f}}(t,u,v)=\{\widehat{f}^{[1]}(t,u,v),\ldots,\widehat{f}^{[N]}(t,u,v)\},\tag{2.3}$$

$$\widehat{f}^{[i]}(t,u,v)=\begin{cases}D^{[i]}X(t^{[i]},t)f(t,u,v), & t<t^{[i]}\\ 0, & t^{[i]}\le t\end{cases}.$$

As earlier, values u and v are constrained by (1.2). Due to (1.3), the saddle point condition in the "small game" is satisfied for system (1.1), (2.2). A starting functional state $\widehat{\mathbf{z}}[t_*^0]=(x[t_*^0],\widehat{\mathbf{x}}[t_*^0])$ for system (1.1), (2.2) is determined by the starting state $x[t_*^0]$ of system (1.1) through (2.1) where $t=t_*^0$.

We introduce a quality index $\widehat{\gamma}$ for motions $\widehat{\mathbf{z}}[t_*^0[\cdot]\vartheta]$ of system (1.1), (2.2) as follows:

$$\widehat{\gamma} = \widehat{\gamma}(\widehat{\mathbf{z}}[\vartheta]) = \mu(\widehat{\mathbf{x}}[\vartheta]). \tag{2.4}$$

Here $\mu(\cdot)$ is the same norm as in (1.4). A value of $\widehat{\gamma}$ given by (2.4) coincides with a value of γ given by (1.4).

Let us consider the differential game (1.1), (2.2)–(2.4) in the space of functional positions $\{t, \widehat{\mathbf{z}}[t]\}$; note that $\widehat{\gamma}$, the payoff in this game, is terminal. This game has a value $\widehat{\rho}^o(t_*, \widehat{\mathbf{z}}[t_*])$ and a saddle point $\{\widehat{u}^o(t, \widehat{\mathbf{z}}[t], \epsilon),\ \widehat{v}^o(t, \widehat{\mathbf{z}}[t], \epsilon)\}$ (see [2]). Here $\widehat{\mathbf{z}}[t_*]$ is an initial state of system (1.1), (2.2), and $\widehat{\mathbf{z}}[t]$ is a current state of this system. The optimal strategies $\widehat{u}^o(t, \widehat{\mathbf{z}}[t], \epsilon)$ and $\widehat{v}^o(t, \widehat{\mathbf{z}}[t], \epsilon)$ can be constructed as extremal ones with respect to the function $\widehat{\rho}^o(t, \widehat{\mathbf{z}}[t])$ (see [1], [2]).

It follows from (1.1)–(1.4) and (2.1)–(2.4) that the value $\widehat{\rho}^o(t, \widehat{\mathbf{z}}[t])$ in the game (1.1), (2.2)–(2.4) coincides with the value $\rho^o(x[t_*^0[\cdot]t])$ in the original game (1.1)–(1.4), and, under condition (2.1), optimal strategies in the game (1.1), (2.2)–(2.4) determine actions u and v optimal in the original game (1.1)–(1.4). Thus, the games (1.1)–(1.4) and (1.1), (2.2)–(2.4) are equivalent. Therefore, solution procedures suggested in [2], [3] for differential games with terminal payoffs can be naturally transformed into those for the original game (1.1)–(1.4) with the history-dependent payoff γ (1.4). One should only take into account that, in contrast to a standard differential system, admissible states $\widehat{\mathbf{z}}[t]$ of system (1.1), (2.2), (2.3) are not arbitrary; their components $x[t]$, $\widehat{x}^{[i]}[t]$, $i = 1, \dots, N$ are subject to the constraints (2.1).

3 Calculation of the Value of the Game

As mentioned earlier, the value $\rho^o(\cdot)$ in the original game (1.1)–(1.4) can be identified through the value $\widehat{\rho}^o(\cdot)$ in the game (1.1), (2.2)–(2.4). For finding $\widehat{\rho}^o(\cdot)$ we use the method of upper convex hulls, [1], [3], related to a more general framework of the program stochastic synthesis, [2].

Take a history $x[t_*^0[\cdot]t_*]$ up to time $t_* \in [t_*^0, \vartheta)$. This history determines a unique functional position $\{t_*, \widehat{\mathbf{z}}[t_*]\}$ (2.1). Choose a partition

$$\Delta_k = \Delta_k\{\tau_j\} = \{\tau_j : \tau_1 = t_*,\ \tau_{j+1} > \tau_j, j = 1, \dots, k,\ \tau_{k+1} = \vartheta\} \tag{3.1}$$

of the segment $[t_*, \vartheta]$. We assume that all $t^{[i]} \in [t_*, \vartheta]$, $i = 1, \dots, N$ from (1.4) are contained in Δ_k.

Introduce a p-dimensional variable

$$\mathbf{l} = \{l^{[1]}, \dots, l^{[N]}\}$$

where $l^{[i]}$ are $p^{[i]}$-dimensional vectors ($i = 1, \dots, N$). Define a domain

$$\mathbf{L} = \{\mathbf{l} : \mu^*(\mathbf{l}) \leq 1\}$$

where $\mu^*(\cdot)$ is the norm dual to $\mu(\cdot)$. Denote

$$\Delta\widehat{\psi}_j(t_*,\mathbf{l}) = \int\limits_{\tau_j}^{\tau_{j+1}} \max_{v\in Q}\min_{u\in P}\langle \mathbf{l}\ ,\ \widehat{\mathbf{f}}(\tau,u,v)\rangle d\tau, \quad j=1,\ldots k. \tag{3.2}$$

Define a recurrent sequence of functions:

$$\begin{aligned} &\widehat{\varphi}_{k+1}(t_*,\mathbf{l}) = 0,\ \widehat{\varphi}_j(t_*,\mathbf{l}) = \{\widehat{\psi}_j(t_*,\cdot)\}^*_{\mathbf{L}}, \\ &\widehat{\psi}_j(t_*,\mathbf{l}) = \Delta\widehat{\psi}_j(t_*,\mathbf{l}) + \widehat{\varphi}_{j+1}(t_*,\mathbf{l}), \\ &\mathbf{l}\in\mathbf{L},\ j=k,k-1,\ldots,1. \end{aligned} \tag{3.3}$$

Here $\{\psi(\cdot)\}^*_{\mathbf{L}}$ denotes the upper convex hull of $\psi(\mathbf{l})$ in the domain $\mathbf{L}$, i.e. the minimal concave function that estimates the function $\psi(\mathbf{l}),\mathbf{l}\in\mathbf{L}$ from above.

Define a value

$$\widehat{e}(t_*,\widehat{\mathbf{z}}[t_*];\Delta_k) = \max_{\mathbf{l}\in\mathbf{L}}[\langle\mathbf{l},\widehat{x}[t_*]\rangle + \widehat{\varphi}_1(t_*,\mathbf{l})]. \tag{3.4}$$

The following theorem is true.

Theorem 3.1 *For every admissible history $x[t^0_*[\cdot]t_*]$ of system (1.1) and every sequence of partitions $\Delta_k\{\tau_j\}$ (3.1) with steps $\delta_k = \max_j(\tau_{j+1}-\tau_j)$ $(k=1,2,\ldots)$, $\lim_{k\to\infty}\delta_k = 0$, it holds that*

$$\lim_{k\to\infty}\widehat{e}(t_*,\widehat{\mathbf{z}}[t_*];\Delta_k) = \widehat{\rho}^o(t_*,\widehat{\mathbf{z}}[t_*]) = \rho^o(x[t^0_*[\cdot]t_*]). \tag{3.5}$$

The first equality in (3.5) follows from a corresponding result for differential games with terminal quality indexes (see [2], [3]). The second equality follows from the equivalency of the games (1.1)–(1.4) and (1.1), (2.2)–(2.4). (Here we recall that the history $x[t^0_*[\cdot]t_*]$ determines a unique functional position $\{t_*,\widehat{\mathbf{z}}[t_*]\}$ through (2.1) where $t=t_*$.)

Remark Theorem 3.1 can be proved directly on the basis of (2.1) and procedure (3.1)–(3.4).

Thus, a procedure of computing the value in the original game is reduced to constructing $\widehat{\varphi}_j(t_*,\mathbf{l})$, the upper convex hulls for $\widehat{\psi}_j(t_*,\mathbf{l})$ in the domain $\mathbf{L}$ of p-dimensional vectors $\mathbf{l}$, $p=\sum_{i=1}^N p^{[i]}$.

It should be noted that N and, hence, p are typically large enough. Therefore, apart of exceptional cases where one can provide efficient algorithms for computing the upper convex hulls $\widehat{\varphi}_j(\cdot)$, practical calculations are rather difficult even if the dimension n of the state vector x of the original system (1.1) is not very high. In this connection it is important that in fact the calculation of value $\rho^0(\cdot)$ through functions $\widehat{\varphi}_j(\cdot)$ defined on a multidimensional space can be reduced to operations in spaces of essentially smaller dimensions. We present a reduction method in the next section.

4 Reduction in Calculation of the Value of the Game

Taking into account (2.1) and (2.3), we write:

$$\langle \mathbf{l}\,,\ \widehat{\mathbf{x}}[t_*]\rangle = \sum_{i=1}^{g(t_*)} \Big\langle l^{[i]},\ D^{[i]}x[t^{[i]}]\Big\rangle + \Big\langle \sum_{i=g(t_*)+1}^{N} X^T\Big(t^{[i]},\vartheta\Big)D^{[i]T}l^{[i]},\ X(\vartheta,t_*)x[t_*]\Big\rangle,$$

$$\Delta\widehat{\psi}_j(t_*,\mathbf{l}) = \int_{\tau_j}^{\tau_{j+1}} \max_{v\in Q}\ \min_{u\in P}\Big\langle \sum_{i=g(\tau_j)+1}^{N} X^T\Big(t^{[i]},\vartheta\Big)D^{[i]T}l^{[i]},\ X(\vartheta,\tau)f(\tau,u,v)\Big\rangle d\tau;$$

here the upper symbol T denotes transposition, $\Delta\widehat{\psi}_j(t_*,\mathbf{l})$ $(j=1,\ldots k)$ are defined by (3.2), and

$$g(t) = \max\Big\{i\ :\ t^{[i]} \le t\Big\},\quad i=1,\ldots,N;$$

if there is no i $(i=1,\ldots,N)$ such that $t^{[i]} \le t$, we set $g(t)=0$.

Therefore, instead of functions $\Delta\widehat{\psi}_j(t_*,\mathbf{l})$, $\widehat{\psi}_j(t_*,\mathbf{l})$ and $\widehat{\varphi}_j(t_*,\mathbf{l})$ of a multi-dimensional vector $\mathbf{l}=\{l^{[1]},\ldots,l^{[N]}\}$, we can deal with appropriate functions of an n-dimensional vector

$$m = \sum_{i=g(\tau_j)+1}^{N} X^T\Big(t^{[i]},\vartheta\Big)D^{[i]T}l^{[i]}$$

and vectors $l^{[i]}, i=1,\ldots,g(\tau_j)$ which constitute only a part of the components of vector $\mathbf{l}$. Let us describe a resulting procedure. The latter is associated with a partition $\Delta_k\{\tau_j\}$ (3.1).

Denote

$$\Delta\psi_j(t_*,m) = \int_{\tau_j}^{\tau_{j+1}} \max_{v\in Q}\ \min_{u\in P}\langle m,\ X(\vartheta,\tau)f(\tau,u,v)\rangle d\tau, \qquad j=1,\ldots,k. \tag{4.1}$$

Introduce domains

$$G_j(t_*) = \{(\mathbf{l}_{(j)} = \{l^{[1]},\ldots,l^{[g(\tau_j)]}\},m) :\ m = \sum_{i=g(\tau_j)+1}^{N} X^T(t^{[i]},\vartheta)D^{[i]T}l^{[i]},\ \mathbf{l}=\{l^{[1]},\ldots,l^{[N]}\}\in\mathbf{L}\} \qquad j=1,\ldots,k. \tag{4.2}$$

Note that for every $j (j = k, k-1, \ldots, 1)$ we have either $g(\tau_{j+1}) = g(\tau_j)$, or $g(\tau_{j+1}) = g(\tau_j) + 1$ because instants $t^{[i]}$ belong to $\Delta_k\{\tau_j\}$ (3.1). Define a recurrent sequence of functions

$$\begin{aligned} \varphi_k(t_*, \mathbf{l}_{(k)}, m) &= \{\Delta\psi_k(t_*, \cdot)\}^*_{G_k(t_*)}, \; (\mathbf{l}_{(k)}, m) \in G_k(t_*), \\ \varphi_j(t_*, \mathbf{l}_{(j)}, m) &= \{\psi_j(t_*, \cdot)\}^*_{G_j(t_*)}, \; (\mathbf{l}_{(j)}, m) \in G_j(t_*), \\ j &= k-1, \ldots, 1, \end{aligned} \tag{4.3}$$

where in the case of $g(\tau_{j+1}) = g(\tau_j)$

$$\psi_j(t_*, \mathbf{l}_{(j)}, m) = \Delta\psi_j(t_*, m) + \varphi_{j+1}(t_*, \mathbf{l}_{(j)}, m)$$

and in the case of $g(\tau_{j+1}) = g(\tau_j) + 1$

$$\psi_j(t_*, \mathbf{l}_{(j)}, m) = \Delta\psi_j(t_*, m) + \max_{l^{[g]}, m_*} \varphi_{j+1}(t_*, \{\mathbf{l}_{(j)}, l^{[g]}\}, m_*),$$

$$(\{\mathbf{l}_{(j)}, l^{[g]}\}, m_*) \in G_{j+1}(t_*), \quad m_* + X^T(t^{[g]}, \vartheta) D^{[g]T} l^{[g]} = m, \quad g = g(\tau_{j+1}).$$

In (4.3) symbol $\{\psi_j(\cdot)\}^*_{G_j}$ denotes the upper convex hull of $\psi_j(\mathbf{l}_{(j)}, m)$ in the domain G_j.

Let

$$e(x[t_*^0[\cdot]t_*]; \Delta_k) = \max_{(\mathbf{l}_{(1)}, m) \in G_1(t_*)} \left[\sum_{i=1}^{g(t_*)} \left\langle l^{[i]}, \; D^{[i]} x[t^{[i]}] \right\rangle + \right. \\ \left. \langle m, \; X(\vartheta, t_*) x[t_*] \rangle + \varphi_1(t_*, \mathbf{l}_{(1)}, m) \right]. \tag{4.4}$$

The following relations between functions $\widehat{\varphi}_j(\cdot)$ constructed by (3.2), (3.3) and functions $\varphi_j(\cdot)$ constructed by (4.1)–(4.3) can be proved. For every j $(j = k, k-1, \ldots, 1)$ and every $(\mathbf{l}_{(j)}, m) \in G_j(t_*)$ we have

$$\varphi_j(t_*, \mathbf{l}_{(j)}, m) = \max_{l^{[g(\tau_j)+1]}, \ldots, l^{[N]}} \widehat{\varphi}_j(t_*, \{\mathbf{l}_{(j)}, l^{[g(\tau_j)+1]}, \ldots, l^{[N]}\})$$

where the maximum is taken under constraints

$$\{\mathbf{l}_{(j)}, l^{[g(\tau_j)+1]}, \ldots, l^{[N]}\} \in \mathbf{L}, \qquad \sum_{i=g(\tau_j)+1}^{N} X^T(t^{[i]}, \vartheta) D^{[i]T} l^{[i]} = m.$$

Hence, comparing (3.4) and (4.4) under the condition (2.1), we obtain

$$\widehat{e}(t_*, \widehat{\mathbf{z}}[t_*]; \Delta_k) = e(x[t_*^0[\cdot]t_*]; \Delta_k)$$

for every history $x[t_*^0[\cdot]t_*]$ and every partition $\Delta_k\{\tau_j\}$ (3.1).

In view of Theorem 3.1 we come to the following.

Theorem 4.1 *For every admissible history $x[t_*^0[\cdot]t_*]$ of system (1.1) and every sequence of partitions $\Delta_k\{\tau_j\}$ (3.1) with steps $\delta_k = \max_j(\tau_{j+1} - \tau_j)$ $(k = 1, 2, \ldots)$, $\lim_{k\to\infty} \delta_k = 0$, it holds that*

$$\lim_{k\to\infty} e(x[t_*^0[\cdot]t_*]; \Delta_k) = \rho^o(x[t_*^0[\cdot]t_*]).$$

Remark Theorem 4.1 can be proved directly on the basis of procedure (4.1)–(4.4).

Thus, the problem is reduced to the construction of functions $\varphi_j(t_*, \mathbf{l}_{(j)}, m)$, the upper convex hulls of $\psi_j(t_*, \mathbf{l}_{(j)}, m)$ in domains $G_j(t_*)$ with respect to the aggregated argument $(\mathbf{l}_{(j)}, m)$. Note that as j approaches 1, the dimension of $(\mathbf{l}_{(j)}, m)$ becomes essentially smaller than p, the dimension of vectors $\mathbf{l} \in \mathbf{L}$ in procedure (3.2)–(3.4).

5 Example

To illustrate the constructions described in the previous sections we consider a particular, relatively simple, control problem (1.1)–(1.4).

Let a motion of a controlled object be described by a scalar function $r[t]$ satisfying the differential equation

$$d^2r/dt^2 = \alpha[t]dr/dt + \beta[t]r + f[t]u + h[t]v, \quad t_*^0 \le t \le \vartheta \tag{5.1}$$

where

$$|\, u \,| \le 1, \quad |\, v \,| \le 1. \tag{5.2}$$

Define a quality index by

$$\gamma = \gamma(r[t_*^0[\cdot]\vartheta]) = \max\{|r[t^{[1]}]|, |r[t^{[2]}]|\} + |r[t^{[3]}]|. \tag{5.3}$$

We assume that the continuous functions $f[t]$ and $h[t]$ are such that

$$\begin{gathered} f[t] > h[t] \ge 0, \quad t_*^0 \le t < \tilde{t} < \vartheta, \\ h[t] > f[t] \ge 0, \quad \tilde{t} < t \le \vartheta, \\ f[\tilde{t}\,] = h[\tilde{t}\,]. \end{gathered} \tag{5.4}$$

Let instants $t^{[i]}$, $i = 1, 2, 3$ in (5.3) satisfy

$$t_*^0 = t^{[1]} < t^{[2]} = \tilde{t} < t^{[3]} = \vartheta. \tag{5.5}$$

We set $x_1 = r$, $x_2 = dr/dt$. Equation (5.1) is equivalent to the following system:

$$\begin{cases} dx_1/dt = x_2 \\ dx_2/dt = \beta[t]x_1 + \alpha[t]x_2 + f[t]u + h[t]v \end{cases} \tag{5.6}$$

Let $X(t, \tau)$ be the fundamental matrix for the homogeneous part of system (5.6):

$$X(t,\tau) = \begin{bmatrix} x_{11}(t,\tau) & x_{12}(t,\tau) \\ x_{21}(t,\tau) & x_{22}(t,\tau) \end{bmatrix}. \tag{5.7}$$

Matrices $D^{[i]}$, $i = 1, 2, 3$ are defined as follows:

$$D^{[i]} = D = [\,1 \quad 0\,]. \tag{5.8}$$

Following the constructions described in the previous sections we obtain the next expressions for $\rho(x[t_*^0[\cdot]t])$, the value of the game (5.1)–(5.8):

$$\begin{aligned} \rho^o(x[t_*^0[\cdot]t]) &= \max\{|x_1[t^{[1]}]|, |x_1[t^{[2]}]|\}+ \\ &+|x_{11}(t^{[3]},t)x_1[t] + x_{12}(t^{[3]},t)x_2[t]| + \int_t^{\vartheta} |x_{12}(t^{[3]},\tau)|(h[\tau] - f[\tau])d\tau \end{aligned} \tag{5.9}$$

if $t \in [t^{[2]}, t^{[3]}]$, and

$$\begin{aligned} \rho^o(x[t_*^0[\cdot]t]) = &\max_{\substack{|l^{[1]}|+|l^{[2]}|\leq 1 \\ |l^{[3]}|\leq 1}} \Big[\, l^{[1]}x_1[t^{[1]}]+ \\ &+\sum_{i=2}^{3} l^{[i]}(x_{11}(t^{[i]},t)x_1[t] + x_{12}(t^{[i]},t)x_2[t])+ \\ &+\int_t^{t^{[2]}} |l^{[2]}x_{12}(t^{[2]},\tau) + l^{[3]}x_{12}(t^{[3]},\tau)|(h[\tau] - f[\tau])d\tau \,\Big]+ \\ &+\int_{t^{[2]}}^{\vartheta} |x_{12}(t^{[3]},\tau)|(h[\tau] - f[\tau])d\tau \end{aligned} \tag{5.10}$$

if $t \in [t^{[1]}, t^{[2]}]$. Optimal strategies $u^o(\cdot)$ and $v^o(\cdot)$ are constructed as extremal ones with respect to the value $\rho^o(\cdot)$ (5.9), (5.10). Explicit calculations result in the following expressions for $u^o(\cdot)$ and $v^o(\cdot)$: if $t \in [t^{[2]}, t^{[3]}]$, then

$$u^o(x[t_*^0[\cdot]t], \epsilon) = -\text{sign}\Big[\, l_{0u}^{[3]}(x[t_*^0[\cdot]t], \epsilon)x_{12}(t^{[3]}, t)\Big]$$

where

$$\begin{aligned} l_{0u}^{[3]}(x[t_*^0[\cdot]t], \epsilon) = \text{argmax}\Big\{ &l^{[3]}(x_{11}(t^{[3]},t)x_1[t] + x_{12}(t^{[3]},t)x_2[t])- \\ &-\epsilon\sqrt{(l^{[3]})^2(x_{11}^2(t^{[3]},t) + x_{12}^2(t^{[3]},t)) + 1} \;:\; |l^{[3]}| \leq 1\Big\}, \end{aligned}$$

and

$$v^o(x[t_*^0[\cdot]t], \epsilon) = \text{sign}\Big[l_{0v}^{[3]}(x[t_*^0[\cdot]t], \epsilon)x_{12}(t^{[3]}, t)\Big]$$

where

$$l_{0v}^{[3]}(x[t_*^0[\cdot]t],\epsilon)\in\operatorname{argmax}\Big\{l^{[3]}(x_{11}(t^{[3]},t)x_1[t]+x_{12}(t^{[3]},t)x_2[t])+$$
$$+\epsilon\sqrt{(l^{[3]})^2(x_{11}^2(t^{[3]},t)+x_{12}^2(t^{[3]},t))+1}\ :\ |l^{[3]}|\le 1\Big\};$$

if $t\in\left[t^{[1]},t^{[2]}\right]$, then

$$u^o(x[t_*^0[\cdot]t],\epsilon)=-\operatorname{sign}\left[l_{0u}^{[2]}(x[t_*^0[\cdot]t],\epsilon)x_{12}(t^{[2]},t)+l_{0u}^{[3]}(x[t_*^0[\cdot]t],\epsilon)x_{12}(t^{[3]},t)\right]$$

where

$$\left(l_{0u}^{[1]},l_{0u}^{[2]},l_{0u}^{[3]}\right)(x[t_*^0[\cdot]t],\epsilon)=\operatorname{argmax}\Big\{l^{[1]}x_1\left[t^{[1]}\right]$$
$$+\sum_{i=2}^{3}l^{[i]}\left(x_{11}\left(t^{[i]},t\right)x_1[t]+x_{12}\left(t^{[i]},t\right)x_2[t]\right)$$
$$+\int_t^{t^{[2]}}\left|l^{[2]}x_{12}\left(t^{[2]},\tau\right)+l^{[3]}x_{12}\left(t^{[3]},\tau\right)\right|(h[\tau]-f[\tau])d\tau$$
$$-\epsilon\sqrt{(l^{[1]})^2+\left(\sum_{i=2}^{3}l^{[i]}x_{11}(t^{[i]},t)\right)^2+\left(\sum_{i=2}^{3}l^{[i]}x_{12}(t^{[i]},t)\right)^2+1}\quad:$$
$$|l^{[1]}|+|l^{[2]}|\le 1,\ |l^{[3]}|\le 1\Big\},$$

and

$$v^o(x[t_*^0[\cdot]t],\epsilon)=\operatorname{sign}\left[l_{0v}^{[2]}(x[t_*^0[\cdot]t],\epsilon)x_{12}(t^{[2]},t)+l_{0v}^{[3]}(x[t_*^0[\cdot]t],\epsilon)x_{12}(t^{[3]},t)\right]$$

where

$$\left(l_{0v}^{[1]},l_{0v}^{[2]},l_{0v}^{[3]}\right)(x[t_*^0[\cdot]t],\epsilon)\in\operatorname{argmax}\Big\{l^{[1]}x_1\left[t^{[1]}\right]$$
$$+\sum_{i=2}^{3}l^{[i]}\left(x_{11}\left(t^{[i]},t\right)x_1[t]+x_{12}\left(t^{[i]},t\right)x_2[t]\right)$$
$$+\int_t^{t^{[2]}}\left|l^{[2]}x_{12}\left(t^{[2]},\tau\right)+l^{[3]}x_{12}\left(t^{[3]},\tau\right)\right|(h[\tau]-f[\tau])d\tau$$
$$+\epsilon\sqrt{(l^{[1]})^2+\left(\sum_{i=2}^{3}l^{[i]}x_{11}(t^{[i]},t)\right)^2+\left(\sum_{i=2}^{3}l^{[i]}x_{12}(t^{[i]},t)\right)^2+1}\quad:$$
$$|l^{[1]}|+|l^{[2]}|\le 1,\ |l^{[3]}|\le 1\Big\};$$

Motions of system (5.1)–(5.5) were simulated with the following parameters and initial data:

$$t_*^0 = 0,\ \vartheta = 4,\ f[t] = 4 - t,\ h[t] = 2,\ \alpha[t] = 0.2,\ \beta[t] = -16,$$
$$t^{[1]} = t_*^0 = 0,\ t^{[2]} = \tilde{t} = 2,\ t^{[3]} = \vartheta = 4,$$
$$r[t_*^0] = x_{01} = 0.2,\ dr/dt\,[t_*^0] = x_{02} = 2.$$

The next simulation results were obtained. The value of the game calculated a priori is $\rho^o(t_*^0, x_{01}, x_{02}) = 0.93$. The optimal control u^o and optimal disturbances v^o give $\gamma_{u^o,v^o} = 0.91 \approx \rho^o(t_*^0, x_{01}, x_{02})$. The optimal control u^o and a non-optimal disturbance v give $\gamma_{u^o,v} = 0.32 < \gamma_{u^o,v^o}$. The optimal disturbance v^o and a non-optimal control u provide $\gamma_{u,v^o} = 3.1 > \gamma_{u^o,v^o}$.

References

[1] Krasovskii, A.N. and Krasovskii, N.N. (1995) *Control under Lack of Information.* Birkhauser, USA.
[2] Krasovskii, N.N. (1985) *The Control of a Dynamic System.* Nauka, Moscow (in Russian).
[3] Krasovskii, N.N. and Reshetova, T.N. (1988) On the program synthesis of a guaranteed control. *Problems of Control and Information Theory*, **17**(6), 333–343.

6 RECEDING HORIZON CONTROL FOR THE STABILIZATION OF DISCRETE-TIME UNCERTAIN SYSTEMS WITH BOUNDED CONTROLLERS

ÉVA GYURKOVICS*

Department of Mathematics, Technical University of Budapest, H-1521 Budapest, Hungary

1 Introduction

In the past two decades, a great deal of interest has been devoted to the design of stabilizing controllers for systems with incomplete or uncertain information. These uncertainties often arise as a result of approximation, imprecision, or imperfect knowledge introduced during the modelling procedure. Moreover, realistic processes are frequently subject to extraneous disturbances with known structural properties and with known bounds. Several hundred papers dealing with this subject have appeared in the literature. (We refer to Corless [7], Leitman [24], Dorato and Muscato [11] for overviews and the numerous references therein.) One possible approach is based on a constructive utilization of Lyapunov stability theory.

In a great part of the literature, continuous systems are considered. Stability issues for discrete time uncertain systems without any control constraint have been discussed, e.g. in Corless [6], Corless and Manela [9], Hollot and Arabacioglu [17], Magaña and Żak [25]. Control constraints have been taken into account for linear systems without uncertainties, e.g. in Benzaouia and Burgat [2], Bitsoris [3], Gutman and Cwikel [12], Vassilaki *et al.* [32] (discrete-time case), Gutman and Hagander [13], and Wredenhagen and Belanger [33] (continuous-time case) and for linear continuous-time uncertain systems in Dolphus and Schmitendorf [10], Corless and Leitmann [8], Polak and Yang [31] and for nonlinear discrete-time systems with exponentially stable nominal uncontrolled part in Gyurkovics and Takács [15].

*Work supported in part by the Hungarian National Foundation for Scientific Research, grant no. T014847

This paper aims to relax the requirement of the exponential stability of the nominal uncontrolled part of the system by making use of a version of the receding horizon control method.

The notion of receding horizon control is not new; it goes back to an early publication of Kleinman [20]. The method has since been revisited by many authors. The basic idea of the method is: at any time t and state $x(t)$, the current control is obtained by solving an optimal control problem with an appropriate Lagrange-type criterion over the time interval $[t, t+T]$ subject to the terminal constraint $x(t+T) = 0$, and setting the current control equal to the value of the optimal control at time t. The stabilizing property of this control method for linear time-varying systems has been established by Kwon and Pearson ([22], [23]).

The method has been generalized for nonlinear systems by Chen and Shaw [4], Michalska and Mayne ([28], [29]) (continuous-time systems) and by Keerthi and Gilbert ([18], [19]), Alamir and Bornard [1] (discrete-time systems).

The terminal constraint $x(t+T) = 0$ which can be interpreted as an "infinite final state penalty" has been relaxed in different ways: in Kwon *et al.* [21] the optimization problem has been considered for linear time-varying systems without any terminal constraint but on "large" time horizon, while in Meadows and Rawlings [27] for nonlinear discrete time systems on an infinite time horizon. This approach is not very promising from a practical point of view because of computational difficultes. The second possibility is the use of a finite final state penalty; i.e. instead of solving a Lagrange-type optimization problem with equality constraint a Bolza-type problem without any constraint is to be solved. This approach has been established for linear time-varying systems in Kwon *et al.* [21] for nonlinear systems in Gyurkovics [14] (continuous time) and in Parisini and Zoppoli [30] (discrete time). Besides the computational complexity, the main difficulty of the methods of Kwon *et al.* [21], Gyurkovics [14] is that the final state penalty function has to satisfy a Riccati-, i.e. a Hamilton-Jacobi-Bellman-type inequality. (This is avoided in Parisini and Zoppoli [30] but both the final state penalty and the time horizon have to be large enough.)

The third way of relaxing the equality constraint is to substitute it by an inequality constraint and to define the controller by an intermittent policy. This way is followed by Michalska and Mayne [29]) (nonlinear, continuous-time) and by Polak and Yang [31] for linear time-invariant systems.

It is well known (see e.g. Hollot and Arabacioglu [17] that an important difference between the continuous- and discrete-time uncertain systems is that the latter can only be stabilized when the "size" of the uncertainties is small enough, even if appropriate structural conditions hold.

Not surprisingly the allowable bound of the uncertainty highly depends on the control constraint and on the set over which the stabilization should be performed. In the formulation of the results of this paper, the control constraint is considered to be primarily given; the set of admissible initial states and uncertainties is determined according to this constraint.

2 Problem Formulation

Let us consider a discrete time system whose evolution is described by the equation

$$x(t+1) = f(x(t)) + B(x(t))u(t) + \Delta f(t, x(t), u(t)), \tag{1}$$

where $x(t) \in \mathbf{R}^n$, $u(t) \in \mathbf{R}^m$ are the state and the control, respectively, $f : \mathbf{R}^n \to \mathbf{R}^n$, $B : \mathbf{R}^n \to \mathbf{R}^{n \times m}$ are given continuously differentiable functions $f(0) = 0$, rank $B(x) = m$ for all $x \in \mathbf{R}^n$, $\Delta f : \mathbf{Z} \times \mathbf{R}^n \times \mathbf{R}^m \to \mathbf{R}^n$ is an unknown function, representing the uncertainty, which is assumed to belong to a given class of functions $\mathcal{F}$, i.e. $\Delta f \in \mathcal{F}$. We shall refer to system (1) as the real system and to

$$x(t+1) = f(x(t)) + B(x(t))u(t) \tag{2}$$

as the nominal system.

The control is subject to the constraint

$$\|u(t)\| \leq \rho,$$

where ρ is a given positive number.

Now, a control strategy has to be determined so that the obtained system will be asymptotically stable about the origin, or at least about a certain neighbourhood of the origin.

The idea of the construction is similar to that of Michalska and Mayne [29] inasmuch as an ellipsoidal neighbourhood of the origin is calculated, where an explicitly given feedback is applied, while outside of the ellipsoid an optimization procedure is performed and the obtained control is applied over a certain time interval.

Standard notation is adapted. In particular, the Euclidean norm is denoted by $\|\cdot\|$, the transpose of a vector or a matrix X is denoted by X'. For any positive definite symmetric matrix S, $\|\cdot\|_S$ means the S-norm of a vector, i.e. $\|y\|_S = \sqrt{y'Sy}$, while $\lambda_m(S)$ and $\lambda_M(S)$ denote the minimal and the maximal eigenvalue of the symmetric matrix S. Moreover, $\mathcal{B}_\rho$ denotes the closed ball about the origin with the radius ρ.

3 Feedback in a Neighbourhood of the Origin

In this section, we shall give a neighbourhood of the origin, in which the controller can be given as the sum of two terms: one of them is a linear feedback for exponential stabilization of the nominal part of the system, the other is a nonlinear feedback for counteracting the uncertainty.

In the sequel we need the following assumption.

Assumption A1 The linearized system

$$x(t+1) = A_0 x(t) + B_0 u(t) \tag{3}$$

is controllable, where $A_0 = Df(0)$, $B_0 = B(0)$.

Let us choose the parameters $0 < \bar{\rho} < \rho$, $0 < \beta < 1$ and the positive definite symmetric matrices $Q \in \mathbf{R}^{n \times n}$, $R \in \mathbf{R}^{m \times m}$. Let $P \in \mathbf{R}^{n \times n}$ be the positive definite symmetric solution of

$$1/\beta^2 A_0' P A_0 - P - 1/\beta^2 A_0' P B_0 (R + B_0' P B_0)^{-1} B_0' P A_0 + Q = 0,$$

and let

$$v_L(x) = -Kx, \quad \text{with} \quad K = (R + B_0^T P B_0)^{-1} B_0^T P A_0. \tag{4}$$

It is well known that system (3) with $u(t) = v_L(x(t))$ is exponentially stable. Since it is the linearization of the system

$$x(t+1) = \bar{f}(x(t)) := f(x(t)) + B(x(t)) v_L(x(t)), \tag{5}$$

it can easily be seen that system (5) is also exponentially stable in a neighbourhood of the origin. In fact, with the notation $A = A_0 - B_0 K$,

$$\begin{aligned}
&\|x(t+1)\|_P^2 - \|x(t)\|_P^2 \\
&\quad = \|Ax(t)\|_P^2 - \|x(t)\|_P^2 + \|\bar{f}(x(t)) - Ax(t)\|_P^2 + 2(\bar{f}(x(t)) - Ax(t)) P A x(t) \\
&\quad \le -(1-\beta^2)\|x(t)\|_P^2 - x'(t)(\beta^2 Q + K' R K) x(t) + \|\bar{f}(x(t)) - Ax(t)\|_P^2 + \\
&\qquad + 2\|Ax(t)\|_P \, \|\bar{f}(x(t)) - Ax(t)\|_P.
\end{aligned}$$

Therefore,

$$\|x(t+1)\|_P^2 - \|x(t)\|_P^2 \le -(1-\beta^2)\|x(t)\|_P^2, \tag{6}$$

if

$$\frac{\|\bar{f}(x(t)) - Ax(t)\|_P}{\|x(t)\|_P} \left(\frac{\|\bar{f}(x(t)) - Ax(t)\|_P}{\|x(t)\|_P} + 2 \frac{\|Ax(t)\|_P}{\|x(t)\|_P} \right) \le \beta^2 \frac{x'(t) Q x(t)}{x'(t) P x(t)}. \tag{7}$$

But inequality (7) surely holds if $\|x(t)\|_P$ is small enough, since $\|\bar{f}(x) - Ax\|_P / \|x\|_P \to 0$ as $x \to 0$, and $\|Ax\|_P / \|x\|_P \le \beta$, while $(x'Qx)/(x'Px) \ge \lambda_m(Q)/\lambda_M(P)$. It is known (see e.g. Gyurkovics and Takács [15]) that condition (6) ensures the exponential stability of (5). To find the appropriate neighbourhood, we shall solve problem $\mathcal{P}_\alpha$ to be defined below.

Let $\lambda_M(K'K, P)$ denote the maximal generalized eigenvalue of the generalized eigenvalue problem $K'Ky = \lambda P y$ and let $\tilde{\rho} = \rho - \bar{\rho}$. If

$$\alpha^2 \le \frac{\tilde{\rho}^2}{\lambda_M(K'K, P)} =: \bar{\alpha}^2,$$

then for any $x \in \mathcal{E}_\alpha = \{x \in \mathbf{R}^n : \|x\|_P \le \alpha\}$, the feedback $v_L(x)$ given in (4) satisfies the constraint

$$\|v_L(x)\| \le \tilde{\rho}.$$

Now $\mathcal{P}_\alpha$ is defined as

$$\mathcal{P}_\alpha : \max\Big\{\alpha \in (0,\bar{\alpha}] : \|\bar{f}(x)\|_P^2 - \beta^2\|x\|_P^2 \leq 0, x \in \mathcal{E}_\alpha\Big\}.$$

According to the considerations above, $\mathcal{P}_\alpha$ has a positive solution α_1.

Let us fix an α, $0 < \alpha \leq \alpha_1$ and consider in $\mathcal{E}_\alpha$ system

$$x(t+1) = \bar{f}(x(t)) + B(x(t))u(t) + \Delta f(t, x(t), -Kx(t) + u(t)) \tag{8}$$

with the control constraint

$$\|u(t)\| \leq \bar{\rho}.$$

We observe that, because of the full rank of $B(x)$ and the continuity of $B(.)$, there exists a $\underline{\lambda} > 0$ such that for all $x \in \mathcal{E}_\alpha$,

$$\lambda_m(S(x)) \geq \underline{\lambda}^2$$

where $S(x) := B^T(x)PB(x)$.

Concerning system (8), the following assumptions will be imposed.

Assumption A2 There exist a function $e : \mathbf{Z} \times \mathcal{E}_\alpha \times \mathcal{B}_{\bar{\rho}}$ and nonnegative constants k_1, k_2, k_3 such that for any $(t, x, u) \in \mathbf{Z} \times \mathcal{E}_\alpha \times \mathcal{B}_{\bar{\rho}}$,

$$\Delta f(t, x, -Kx + u) = B(x)e(t, x, u),$$

$$\|e(t, x, u)\|_{S(x)} \leq k_1\|x\|_P + k_2\|u\|_{S(x)} + k_3,$$

where $k_2 < 1$.

Assumption A3

$$a^2 < 1 - \beta^2 \quad \text{and} \quad b < \underline{\lambda}\bar{\rho},$$

where

$$a = \frac{k_1}{1 - k_2}, \qquad b = \frac{k_3}{1 - k_2}.$$

Remark Uncertainties satisfying Assumption A2 are said to be matched and cone-bounded (see e.g. Corless and Manela [9]). The condition $k_2 < 1$ seems to be reasonable, since $e(t, x, u) = \tilde{k}_2(t)u$ with an unknown $\tilde{k}_2$ is a possible realization of the uncertainty, therefore in the case of $|\tilde{k}_2(t)| \geq 1$, the uncertainty might completely destroy the intended effect of the control.

We introduce the following notation

$$\gamma(x) = a\|x\|_P + b, \quad \mu(x) = B^T(x)P\bar{f}(x), \quad \rho(x) = \sqrt{\lambda_m(S(x))}\bar{\rho},$$

and

$$\mathcal{D} = \{x \in \mathbf{R}^m : \gamma(x) < \underline{\lambda}\bar{\rho}\}.$$

Because of Assumption A3, $0 \in \mathcal{D}$, therefore there exists an $\alpha' > 0$ such that

$$\mathcal{E}_{\alpha'} \subset \mathcal{D} \cap \mathcal{E}_{\alpha}$$

Remark If a and b are small enough to satisfy the inequality

$$a\alpha + b < \underline{\lambda}\bar{\rho}$$

then $\mathcal{D} \cap \mathcal{E}_{\alpha} = \mathcal{E}_{\alpha}$, thus $\alpha' = \alpha$ can be taken.

From the definitions of the considered quantities it follows that

$$\gamma(x) \le \rho(x), \quad \text{if} \quad x \in \mathcal{E}_{\alpha'},$$

thus the proposed controller can be defined as

$$p : \mathcal{E}_{\alpha'} \to \mathcal{B}_{\bar{\rho}},$$

$$p(x) = \begin{cases} -\rho(x)\dfrac{S^{-1}(x)\mu(x)}{\|\mu(x)\|_{S^{-1}(x)}}, & \text{if C1: } \rho(x) < \|\mu(x)\|_{S^{-1}(x)}, \\ -\gamma(x)\dfrac{S^{-1}(x)\mu(x)}{\|\mu(x)\|_{S^{-1}(x)}}, & \text{if C2: } \gamma(x) < \|\mu(x)\|_{S^{-1}(x)} \le \rho(x), \\ -S^{-1}(x)\mu(x), & \text{if C3: } \|\mu(x)\|_{S^{-1}(x)} \le \gamma(x). \end{cases} \tag{9}$$

The desired feedback is $u = -Kx + p(x)$, and this results in the closed-loop system

$$x(t+1) = \bar{f}(x(t)) + B(x(t))(p(x(t)) + e(t, x(t), p(x(t)))). \tag{10}$$

Theorem 1 *Consider the uncertain system (8) satisfying Assumptions A1–A3, subject to control $p(\cdot)$ given by (9). If*

$$W = b\frac{a + \sqrt{1-\beta^2}}{1 - \beta^2 - a^2} < \alpha'$$

and $\bar{\beta}$ is such a constant that $a^2 + \beta^2 < \bar{\beta}^2 < 1$, then the closed-loop system (10) is exponentially stable about $\mathcal{E}_W$ with rate $\bar{\beta}$ and with region of attraction $\mathcal{E}_{\alpha'}$.

Corollary Under the conditions of Theorem 1, if $\Delta f(t, 0, 0) = 0$ (i.e. $b = 0$) then the closed-loop system (10) is exponentially stable about the origin with a rate $\bar{\beta}$, where $a^2 + \beta^2 < \bar{\beta}^2 < 1$ and with region of attraction $\mathcal{E}_{\alpha'}$.

The proof of Theorem 1 can be found in Gyurkovics and Takács [16].

Remark It is worth noting that the parameters of the controller discussed above can be determined off-line. Though there is no systematic way for finding the "best" $\bar{\rho}$, K, $\mathcal{E}_{\alpha}$, several numerical experiments may be peformed by changing the parameters β, $\bar{\rho}$, Q, R to calculate the "most satisfactory" control and region of attraction.

4 Control of the Uncertain System Outside of $\mathcal{E}_{\alpha'}$

If the initial point x_0 is outside of $\mathcal{E}_{\alpha'}$, then either $\|Kx_0\| > \tilde{\rho}$ or $x'Px$ is not a suitable Lyapunov function for system (5), thus the linear feedback may be either not admissible or not stabilizing for the nominal system, therefore the Lyapunov approach for the stabilization of the uncertain system is not applicable.

To extend the domain in which system (1) with an appropriate controller is asymptotically stable about a set — specifying more exactly, about $\mathcal{E}_W$ — we may proceed as follows. Consider the set of initial points that can be steered along the nominal system (2) into an ellipsoid $\mathcal{E}_{\alpha''}$, with $\alpha'' < \alpha'$, within a certain number of steps. For such an initial state we solve — at least approximately — an optimal control problem over the time interval $[t_0, t_f] \subset \mathbf{Z}$ for the *nominal system* with respect to a cost criterion and to the inequality type final state constraint $x(t_f) \in \mathcal{E}_{\alpha''}$, and apply the obtained control to the *real system* over the interval $[t_0, t_0 + l_0] \subset \mathbf{Z}$, $t_0 + l_0 \leq t_f$. The parameters of the controller and the bounds of the admissible uncertainty have to be determined so that the trajectory of the nominal system starting from the current state of the real system and corresponding to the restriction of the obtained control sequence to $[t_0 + l_0, t_f]$ should enter into of $\mathcal{E}_{\alpha'}$. Afterwards the current state of the real system is taken as the initial one, and the calculations above are repeated. This receding horizon type controller is completed with an additional term that should reduce the deviation of the real trajectory from the nominal one. The procedure is iterated until the trajectory of the real system enters into the set $\mathcal{E}_{\alpha'}$ and afterwards the feedback described in the previous section is applied.

Let us choose again a parameter $\bar{\rho}$, $0 < \bar{\rho} < \rho$, and let us denote by $\tilde{\rho} = \rho - \bar{\rho}$.

Let $U(t_0, t_f)$ denote the control sequence $U(t_0, t_f) = \{u(t_0), u(t_0 + 1), \ldots, u(t_f - 1)\}$ such that $\|u(i)\| < \tilde{\rho}$, $i = t_0, \ldots, t_f - 1$ and let $x^{U(t_0,t_f)}(.; t_0, x_0)$ and $x_r^{U(t_0,t_f)}(.; t_0, x_0)$ denote the solution of the nominal and the real system due to the control sequence $U(t_0, t_f)$ and satisfying the initial conditions $x^{U(t_0,t_f)}(t_0; t_0, x_0) = x_0$ and $x_r^{U(t_0,t_f)}(t_0; t_0, x_0) = x_0$, respectively. Let $\mathcal{X}$ be the set of all initial states $x_0 \in \mathbf{R}^n$, for which there exists a control sequence $U(0, T)$ such that for the corresponding solution of the nominal system $x^{U(0,T)}(T; 0, x_0) \in \mathcal{E}_{\alpha''}$ holds, where $T \leq T_{max}$ with a given upper bound T_{max} on the admissible time horizon.

If $\mathcal{X} \backslash \mathcal{E}_{\alpha'} = \emptyset$ then the controller of the previous section suffices and the stabilization has to be performed over the set $\mathcal{E}_{\alpha'}$.

Suppose that $\mathcal{X} \backslash \mathcal{E}_{\alpha'} \neq \emptyset$. For any $x \in \mathcal{X} \backslash \mathcal{E}_{\alpha'}$ we define the set of feasible pairs $(T, U(0, T))$ as

$$Z(x) = \{(T, U(0, T))\colon U(0, T) = \{u_0, u_1, \ldots, u_{T-1}\}, \quad \|u_j\| \leq \tilde{\rho},$$

$$T \in \mathbf{N}, \quad T \leq T_{max}, \quad x^U(T; 0, x) \in \mathcal{E}_{\alpha''}, \quad x^U(i; 0, x) \notin \mathcal{E}_{\alpha''}, \quad 0 \leq i < T\} \tag{11}$$

Let the nominal system (2) be subject to the cost function

$$J(t_0, t_f, U(t_0, t_f), x(0)) = \sum_{j=t_0}^{t_f-1} [x(j)'Qx(j) + u'(j)Ru(j)],$$

where Q and R are given positive definite symmetric matrices of appropriate size. Consider the free end-time optimization problem

$$\mathcal{P}(x) : \inf_{Z(x)} J(0, T, U(0, T), x).$$

First we shall define the receding horizon controller whithout the additional control term. For simplicity, we shall denote by x_{t_j} and $x^r_{t_j}$ the state of the nominal system and the real system at the moment t_j, i.e. $x_{t_j} = x^{U_j}(t_j; t_{j-1}, x^r_{t_{j-1}})$ and $x^r_{t_j} = x_r^{U_j}(t_j; t_{j-1}, x^r_{t_{j-1}})$, where $U_j = U_j(0, T_j)$ is the control obtained in the course of the following algorithm. This algorithm is a disrete-time version of the "Robust Receding Horizon Controller" of Michalska and Mayne [29].

Algorithm Let $x_0 \in \mathcal{X}$; fix $l > 0$, $l \in \mathbf{N}$.

Step 0. Let $t_0 = 0$.

a) If $x_0 \in \mathcal{E}_{\alpha'}$, then employ $v_L + p$.

b) If $x_0 \in \mathcal{X} \backslash \mathcal{E}_{\alpha'}$, then find a feasible pair $(T_0, U_0(0, T_0)) \in Z(x_0)$ for problem $\mathcal{P}(x_0)$ and apply the control $u(i) = u_i$, $i = 0, \ldots, l_0 - 1$, to the real system, where $l_0 = \min\{T_0, l\}$, and let $t_1 = l_0$.

Step j. a) If $x^r_{t_j} \in \mathcal{E}_{\alpha'}$, then employ $v_L + p$.

b) (i) If $x^r_{t_j} \in \mathcal{X} \backslash \mathcal{E}_{\alpha'}$ then compute a feasible pair $(T_j', U'(0, T_j')) \in Z(x^r_{t_j})$ in the following way. Let $\tilde{U}$ be the restriction of $U_{j-1}(0, T_{j-1})$ to $l_{j-1}, \ldots, T_{j-1}$, i.e. $\tilde{U} = \{u_{l_{j-1}}, \ldots, u_{T_{j-1}}\}$ and find

$$z_j = x^{\tilde{U}}(T_{j-1}; l_{j-1}, x^r_{t_j}) \in \mathcal{E}_{\alpha'}.$$

Starting with the initial state z_j, compute the control and state pairs $(u_L(\cdot), x_L(\cdot))$ for the nominal system with the controller v_L until the trajectory enters to $\mathcal{E}_{\alpha''}$. Let τ_j be such that $x_L(\tau_j) \in \mathcal{E}_{\alpha''}$, and $x_L(i) \notin \mathcal{E}_{\alpha''}$ for $0 \leq i < \tau_j$. Let $T_j' = T_{j-1} - l_{j-1} + \tau_j$ and let $U'(0, T_j') = \{u_0', \ldots, u'_{T_j'}\}$ be defined by

$$u_k' = \begin{cases} u_{l_{j-1}+k}, & k = 0, 1, \ldots, T_{j-1} - l_{j-1} - 1, \\ u_L(k - (T_{j-1} - l_{j-1})), & k = T_{j-1} - l_{j-1}, \ldots, T_j' - 1. \end{cases}$$

(ii) Compute a feasible pair $(T_j, U_j(0, T_j)) \in Z(x^r_{t_j})$ for which

$$J(t_j, t_j + T_j, U(0, T_j), x^r_{t_j}) \leq J(t_j, t_j + T_j', U'(0, T_j'), x^r_{t_j}).$$

Apply the control $u(t_j + i) = u_i$, $i = 0, \ldots, l_j - 1$ to the real system, where $l_j = \min\{T_j, l\}$, let $t_{j+1} = t_j + l_j$ and let $j = j + 1$.

This algorithm is well-defined if $z_j \in \mathcal{E}_{\alpha'}$ in Step j. In order to ensure this, the following assumption is needed. Let $\mathcal{X}_1 = \mathcal{X} + \mathcal{B}_\epsilon$, where $\epsilon > 0$ is an appropriately chosen constant (see condition (14) below).

Assumption A4 There exists a constant L such that

$\|f(x) + B(x)u - f(\bar{x}) - B(\bar{x})u\|_P \leq L\|x - \bar{x}\|_P$ for any $u \in \mathcal{B}_{\tilde{\rho}}$ and any $x, \bar{x} \in \mathcal{X}_1$.

Remark If $\mathcal{X}_1$ is bounded then Assumption A4 is obviously satisfied.

Observation 1 For any control sequence $U(0, T)$ with $\|u_i\| \leq \tilde{\rho}$ and $T \leq T_{max}$ and for any $x, \bar{x} \in \mathcal{X}$ we have

$$\|x^U(T; 0, x) - x^U(T; 0, \bar{x})\|_P \leq L^T\|x - \bar{x}\|_P.$$

Since $x^{U_{j-1}(0,T_{j-1})}(T_{j-1,}; 0, x^r_{t_{j-1}}) \in \mathcal{E}_{\alpha''}$, we have $z_j \in \mathcal{E}_{\alpha'}$ if

$$\|x^{U(0,T_{j-1})}(l_{j-1}; 0, x^r_{t_{j-1}}) - x^r_{t_j}\|_P \leq \frac{\alpha' - \alpha''}{L^{T_{j-1}-l_{j-1}}}. \tag{12}$$

The left-hand side of (12) depends of course on the size of the uncertainty. In the estimation of the deviation of the states of nominal and real systems, the assumption below is required.

Assumption A5 There exist a function $\bar{e} : \mathbf{Z} \times (\mathcal{X}_1 \backslash \mathcal{E}_{\alpha'}) \times \mathcal{B}_\rho \to \mathbf{R}^m$ and a constant $\bar{k_2}$ such that for any $(t, x, u) \in \mathbf{Z} \times (\mathcal{X}_1 \backslash \mathcal{E}_{\alpha'}) \times \mathcal{B}_\rho$,

$$\Delta f(t, x, u) = B(x)\bar{e}(t, x, u),$$

and

$$\|\bar{e}(t, x, u + v)\|_{S(x)} \leq \|\bar{e}(t, x, u)\|_{S(x)} + \bar{k_2}\|v\|_{S(x)},$$

where $\bar{k_2} < 1$. Furthermore, there exists a bounded function $\kappa : (\mathcal{X}_1 \backslash \mathcal{E}_{\alpha'}) \to \mathbf{R}$ such that for any $(t, x) \in \mathbf{Z} \times (\mathcal{X}_1 \backslash \mathcal{E}_{\alpha'})$

$$\|\bar{e}(t, x, 0)\|_{S(x)} \leq \kappa(x).$$

Observation 2 For any $x \in \mathcal{X}$ and any control sequence $U(0, T)$ with $\|u_i\| \leq \tilde{\rho}$ and $T \leq T_{max}$ we have

$$\|x^{U(0,T)}(T; 0, x) - x_r^{U(0,T)}(T; 0, x)\|_P \leq \sup_{x \in \mathcal{X}_1 \backslash \mathcal{E}_{\alpha'}} \left(\kappa(x) + \bar{k_2}\sqrt{\lambda_M(S(x))}\tilde{\rho}\right)\frac{1 - L^T}{1 - L}.$$

In fact, using the notation $x_i = x^{U(0,T)}(i; 0, x)$ and $x^r_i = x_r^{U(0,T)}(i; 0, x)$, we have

$$\begin{aligned}\|x_{i+1} - x^r_{i+1}\|_P &\leq L\|x_i - x^r_i\|_P + \|B(x^r_i)\bar{e}(k, x^r_i, u_i)\|_P \leq \\ &\leq L\|x_i - x^r_i\|_P + \kappa(x^r_i) + \bar{k_2}\|u_i\|_{S(x^r_i)},\end{aligned} \tag{13}$$

provided that $x_i^r \in \mathcal{X}_1$. Since $x_0 = x_0^r$, the required estimation follows by induction immediately. Note that if

$$\epsilon \geq \varphi(T) := \max\left\{ \sup_{x \in \mathcal{X}_1 \setminus \mathcal{E}_{\alpha'}} \left(\kappa(x) + \bar{k}_2 \sqrt{\lambda_M(S(x))}\tilde{\rho}\right), \sup_{x \in \mathcal{X}_1 \setminus \mathcal{E}_{\alpha'}} \left(\kappa(x) + \bar{k}_2 \sqrt{\lambda_M(S(x))}\tilde{\rho}\right) \frac{1 - L^T}{1 - L} \right\} \tag{14}$$

then $x_i^r \in \mathcal{X}_1$ holds for all $i = 0, 1, \ldots, T$.

Observation 3 In consequence of the inequality (6), an overbound for τ_j suitable in Step j is

$$\bar{\tau} := \left\lceil \frac{\ln \alpha'' - \ln \alpha'}{\ln \beta} \right\rceil + 1, \tag{15}$$

therefore if l is chosen to be $l = \bar{\tau} + 1$, then

$$T_j < T_{j-1}.$$

Observation 4 If for $l = \bar{\tau} + 1$, where $\bar{\tau}$ is given by (15)

$$\varphi(l) \leq (\alpha' - \alpha'')\min\{1, L^{l - T_{max}}\} \tag{16}$$

then $z_j \in \mathcal{E}_{\alpha'}$ holds in Step j.

Thus we may conclude that if Assumptions A1–A5 and the conditions of Observations 1–4 hold, then the Algorithm yields a trajectory of the real system which remains in the set $\mathcal{X}_1$ and which enters into $\mathcal{E}_{\alpha'}$ in finite number of steps not greater than T_{max}.

The result can be summarized in the following theorem.

Theorem 2 *Suppose that assumptions A1–A5 hold, l is chosen according to Observation 3, inequalities (14) and (16) are satisfied. Then system (1) with the controller determined by the Algorithm is exponentially stable about* $\mathcal{E}_W$.

Proof The statement of the theorem immediately follows from Theorem 1 and Observations 1–4.

Consider now, how to modify the receding horizon controller in order to improve its robustness.

Let $\underline{\lambda}^c$ denote the value

$$\underline{\lambda}^c = \inf_{x \in \mathcal{X} \setminus \mathcal{E}_{\alpha'}} \sqrt{\lambda_m(S(x))}.$$

Inequality

$$\sup_{x \in \mathcal{X}_1 \setminus \mathcal{E}_{\alpha'}} \left(\kappa(x) + \bar{k}_2 \sqrt{\lambda_M(S(x))}\tilde{\rho}\right) < \underline{\lambda}^c \bar{\rho}(1 - \bar{k}_2) \tag{17}$$

ensures that the uncertainty does not exceed the effect of the control, so it can be considered as a restriction due to the control constraint. If condition (16) is not too strict in comparison with the restriction (17), then there is no need to use an additional term in the control to reduce the distance between the nominal and the real trajectories.

In what follows, we assume that

$$(\alpha' - \alpha'')\min\{1, L^{l-T_{max}}\} < \underline{\lambda}^c \bar{\rho}(1 - \bar{k}_2) \tag{18}$$

and define the function

$$\begin{aligned}
\gamma^c(x) &= \frac{1}{1-\bar{k}_2}(\kappa(x) + \sqrt{\lambda_M(S(x))}\bar{k}_2\tilde{\rho}), && x \in \mathcal{X}_1 \backslash \mathcal{E}_{\alpha'} \\
\eta(x,\xi,u) &= f(x+\xi) - f(x) + (B(x+\xi) - B(x))u, && x \in \mathcal{X}_1 \backslash \mathcal{E}_{\alpha'}, u \in \mathcal{B}_{\bar{\rho}}, \xi \in \mathcal{E}_{\varphi(l)} \\
\mu^c(x,\xi,u) &= B^T(x)P\eta(x,\xi,u) && x \in \mathcal{X}_1 \backslash \mathcal{E}_{\alpha'}, u \in \mathcal{B}_{\bar{\rho}}, \xi \in \mathcal{E}_{\varphi(l)}
\end{aligned}$$

If $\rho(x)$ is the same as in the previous section, then the assumption

$$\gamma^c(x) \leq \rho(x) \tag{19}$$

is less strict than (17). Thus we may assume (19) to be valid. The proposed additional control term is the following:

$$p^c : \quad (\mathcal{X}_1 \backslash \mathcal{E}_{\alpha'}) \times \mathcal{E}_\nu \times \mathcal{B}_{\bar{\rho}} \to \mathcal{B}_{\bar{\rho}}$$

$$p^c(x,\xi,u) = \begin{cases} -\rho(x+\xi)\dfrac{S^{-1}(x+\xi)\mu^c(x,\xi,u)}{\|\mu^c(x,\xi,u)\|_{S^{-1}(x+\xi)}}, & \text{if C1}': \|\mu^c(x,\xi,u)\|_{S^{-1}(x)} \geq \rho(x+\xi), \\ -\gamma^c(x+\xi)\dfrac{S^{-1}(x+\xi)\mu^c(x,\xi,u)}{\|\mu^c(x,\xi,u)\|_{S^{-1}(x+\xi)}}, & \text{if C2}': \begin{array}{l}\gamma^c(x+\xi) \leq \|\mu^c(x,\xi,u)\|_{S^{-1}(x)} \\ < \rho(x+\xi),\end{array} \\ -S^{-1}(x+\xi)\mu^c(x,\xi,u), & \text{if C3}': \|\mu^c(x,\xi,u)\|_{S^{-1}(x)} < \gamma^c(x+\xi). \end{cases} \tag{20}$$

A straightforward computation shows that, under Assumption A5, estimations

$$\|e(k, x+\xi, u+p^c(x,\xi,u))\|_{S(x+\xi)} \leq \begin{cases} \rho(x+\xi) & \text{if C1}' \\ \gamma^c(x+\xi) & \text{if C2}' \text{ or C3}' \end{cases} \tag{21}$$

and

$$\|p^c(x,\xi,u)\| \leq \bar{\rho}$$

hold true in the domain of definition.

Lemma 1 Suppose that Assumptions A1–A5 and condition (19) hold. Let $(x,\xi,u) \in (\mathcal{X}_1 \backslash \mathcal{E}_{\alpha'}) \times \mathcal{E}_\nu \times \mathcal{B}_{\bar{\rho}}$ and let

$$\begin{aligned}
\bar{x} &= f(x) + B(x)u; \\
\bar{x} + \bar{\xi}(k) &= f(x+\xi) + B(x+\xi)(u + p^c(x,\xi,u) + e(k, x+\xi, u+p^c(x,\xi,u))).
\end{aligned}$$

Then

$$\|\bar{\xi}(k)\|_P^2 \leq \begin{cases} \|\eta(x,\xi,u)\|_P^2 & \text{if C1}' \text{ or C2}', \\ \|\eta(x,\xi,u)\|_P^2 + \|e(k,x+\xi,u+p^c(x,\xi,u))\|_{S(x)}^2 & \text{if C3}'. \\ \qquad - \|\mu^c(x,\xi,u)\|_{S^{-1}(x+\xi)}^2, & \end{cases}$$

Proof We have to estimate the expression

$$\|\bar{\xi}(k)\|_P^2 = \|\eta(x,\xi,u)\|_P^2 + \theta(k,x,\xi,u),$$

where

$$\theta = 2(p^c)'B'P\eta + 2e'B'P\eta + 2e'B'PBp^c + (p^c)'B'PBp^c + e'B'PBe.$$

Using the estimation (21) and the Cauchy-Schwarz inequality, we have that in case C1′

$$\theta \leq 2\left(-\rho\|B'P\eta\|_{S^{-1}} + \frac{\|B'P\eta\|_{S^{-1}} - \rho}{\|B'P\eta\|_{S^{-1}}}\|e\|_S\,\|B'P\eta\|_{S^{-1}} + \rho^2\right) \leq 0,$$

in case C2′

$$\theta \leq 2\left(-\gamma\|B'P\eta\|_{S^{-1}} + \frac{\|B'P\eta\|_{S^{-1}} - \gamma}{\|B'P\eta\|_{S^{-1}}}\|e\|_S\,\|B'P\eta\|_{S^{-1}} + \gamma^2\right) \leq 0,$$

and in case C3′

$$\begin{aligned}\theta &\leq -\|B'P\eta\|_{S^{-1}}^2 + \|e(k,x+\xi,u+p^c(x,\xi,u))\|_{S(x)}^2 = \\ &= \|e(k,x+\xi,u+p^c(x,\xi,u))\|_{S(x)}^2 - \|\mu^c\|_{S^{-1}}^2.\end{aligned}$$

Let us compare the estimation of Lemma 1 with the estimation obtained with $p^c = 0$, i.e. consider

$$\bar{x} + \bar{\xi}_0(k) = f(x+\xi) + B(x+\xi)(u + e(k,x+\xi,u)).$$

Then

$$\|\bar{\xi}_0(k)\|_P^2 \leq \|\eta(x,\xi,u)\|_P^2 + \theta_0(k,x,\xi,u),$$

where

$$\theta_0 = 2e'B'P\eta + e'B'PBe = 2e'S^{1/2}S^{-1/2}B'P\eta + e'B'PBe.$$

Now,

$$\theta_0 \leq 2\|e\|_S\,\|\mu^c\|_{S^{-1}} + \|e\|_S^2,$$

where equality is achieved if $S^{1/2}e$ is parallel with $S^{-1/2}B'P\eta$. Since θ is nonpositive in cases C1′ and C2′, the use of the additional control term provides a better estimation in these cases.

Taking into account that

$$\|e(k, x+\xi, u+p^c(x,\xi,u+p))\|_{S(x+\xi)} \leq$$
$$\|e(k, x+\xi, u)\|_{S(x+\xi)} + \bar{k}_2\|p^c(x,\xi,u+p)\|_{S(x+\xi)}$$

and

$$\|p^c(x,\xi,u)\|_{S(x+\xi)} = \|\mu^c(x,\xi,u)\|_{S^{-1}(x+\xi)} \quad \text{in case C3}',$$

we obtain

$$\left(2\|e\|_S\|\mu^c\|_{S^{-1}} + \|e\|_S^2\right) - \left(\left(\|e\|_S + \bar{k}_2\|\mu^c\|_{S^{-1}}\right)^2 - \|\mu^c\|_{S^{-1}}^2\right) =$$
$$= \left(\|e\|_S^2 + \|\mu^c\|_{S^{-1}}\right)^2 - \left(\|e\|_S + \bar{k}_2\|\mu^c\|_{S^{-1}}\right)^2 \geq 0$$

since $\bar{k}_2 < 1$ and equality is only valid if $\mu^c = 0$; thus, in case C3′, the additional control term gives an estimation not worse than without it.

Now we consider, how to modify the Algorithm.

Assume that at the end of Step j we have the control sequence $U(0, T_j) = \{u_0, \ldots, u_{l_j-1}, u_{l_j}, \ldots, u_{T_j-1}\}$. For $i = 0, 1, \ldots, l_j - 1$ we calculate with $x(t_j) = x^r(t_j) = x^r_{t_j}$

$$x(t_j + i + 1) = f(x(t_j + i)) + B(x(t_j + i))u_i$$

and

$$x^r(t_j + i + 1) = f(x^r(t_j + i)) + B(x^r(t_j + i))[u_i + p^c(x(t_j + i), \xi(t_j + i), u_i)$$
$$+ e(t_j + i, x^r(t_j + i), u_i + p^c(x(t_j + i), \xi(t_j + i), u_i))]$$

where $\xi(t_j + i) := x^r(t_j + i) - x(t_j + i)$, i.e. the proposed controller is

$$u(t_j + i) = u_i + p^c(x(t_j + i), \xi(t_j + i), u_i). \tag{22}$$

According to Lemma 1

$$\|\xi(t_j + i + 1)\|_P^2 \leq \begin{cases} \|\eta(x(t_j + i), \xi(t_j + i), u_i)\|_P^2, & \text{if C1}' \text{ or C2}' \\ \|\eta(x(t_j + i), \xi(t_j + i), u_i)\|_P^2 \\ \quad - \|\mu(x(t_j + i), \xi(t_j + i), u_i)\|_{S^{-1}(x^r(t_j+i))}^2 & \text{if C3}' \\ \quad + \|e(t_j + i, x^r(t_j + i), u_i \\ \quad + p^c(x(t_j + i), \xi(t_j + i), u_i)\|_{S^{-1}(x^r(t_j+i))}^2 \end{cases}$$

If for each $i = 0, 1, \ldots, l_j - 1$, $B'(x^r(t_j + i))P\eta(x(t_j + i), \xi(t_j + i), u_i) = 0$, then the receding horizon controller is not modified, otherwise the estimation of the deviation of real and nominal states is better if the additional control term is applied than if it is not. Therefore, Theorem 2 remains valid for the modified controller (22).

5 Conclusion

We have presented, in this paper, a method of stabilization of nonlinear, discrete-time uncertain systems in which the uncertainties are modeled deterministically rather than stochastically. The control has been subject to the hard constraint $\|u\| \leq \rho$ with a prespecified constant ρ. The nominal system has been stabilized by a receding horizon type controller. This controller has been given in a neighbourhood of the origin in the form of a linear feedback which has ensured the exponential stability of the closed-loop nominal system with a quadratic Lyapunov function. The uncertain system has been stabilized by adding a nonlinear feedback term to the previous controller. The nonlinear feedback has been constructed by using Lyapunov stability theory. Outside of the neighbourhood in question, a receding horizon type open-loop controller has been calculated for the nominal system, and this controller has been applied to the uncertain system in a small number of steps. The robustness of this controller is improved by a feedback constructed from the deviation of real and nominal trajectories. The receding horizon control method has been used until the state of the uncertain system enters the given neighbourhood of the origin.

References

[1] Alamir, M. and Bonard, G. (1994) On the stability of receding horizon control of nonlinear discrete-time systems. *Systems and Control Letters*, **23**, 291–296.

[2] Benzaouia, A. and Burgat, C. (1988) Regulator problem for linear discrete-time systems with non-symmetrical constrained control. *Int. J. Control*, **48**, 2441–2451.

[3] Bitsoris, G. (1988) Positively invariant polyhedral sets of discrete-time linear systems. *Int. J. Control*, **47**, 1713–1726.

[4] Chen, C.C. and Shaw, L. (1982) On receding horizon feedback control. *Automatica*, **18**, 349–352.

[5] Chen, Y.H. (1987) Design of deterministical controllers for uncertain dynamical systems. *Proc. American Control Conference*, Mineapolis, MN, 1259–1263.

[6] Corless, M. (1986) Stabilization of uncertain discrete-time systems. *Proc. IFAC Workshop on Model Error Concepts and Compensation*, Boston

[7] Corless, M. (1993) Control of uncertain nonlinear systems. *ASME J. Dynam. Syst., Measurem., and Control*, **115**, 362–372.

[8] Corless, M. and Leitmann, G. (1993) Bounded controllers for robust exponential convergence. *J. Optimization Theory and Appl.*, **76**, 1–12.

[9] Corless, M. and Manela, J. (1986) Control of uncertain discrete-time systems. *Proc. American Control Conference*, Seattle, WA, 515–520.

[10] Dolphus, R.M. and Schmitendorf, W.E. (1991) Stability analysis for a class of linear controllers under control constraints. *Proc. 30th Conference on Decision and Control*, Brighton, England, Vol. 1., 77–80.

[11] Dorato, P. and Muscato, G. (1993) Bibliography on robust control. *Automatica*, **29**, 201–213.

[12] Gutman, P.O. and Cwikel, M. (1987) An algorithm to find maximal state constraint sets for discrete-time linear dynamical systems with bounded controls and states. *IEEE Trans. Automatic Control*, **AC–32**, 251–254.

[13] Gutman, P.O. and Hagander, P. (1985) A new design of constrained controllers for linear systems. *IEEE Trans. Automatic Control*, **AC–30**, 22–33.

[14] Gyurkovics, É. (1996) Receding horizon control for the stabilization of nonlinear uncertain systems described by differential inclusions. *J. Math. Systems, Estimation, and Control*, **6**(3), 363–366. (summary; full electronic manuscript = 16 pp, retrieval code: 18283).

[15] Gyurkovics, É. and Takács, T. (1994) Exponential convergence of discrete-time systems using constrained control. *PU. M. A*, **5**, 15–25.

[16] Gyurkovics, É. and Takács, T. (1995) Exponential stabilization of discrete-time uncertain systems under control constraints. *Proc. of 3rd European Control Conference*, Rome, Italy, September, 3636–3641.

[17] Hollot, C.V. and Arabacioglu, M. (1987) ℓth-step Lyapunov min-max controllers: stabilizing discrete-time systems under real parameter variations. *Proc. American Control Conference*, Minneapolis, MN, 496–501.

[18] Keerthi, S.S. and Gilbert, E.G. (1987) Moving-horizon approximations for a general class of optimal nonlinear infinite-horizon discrete-time systems. In *Information Sciences and Sytems, Proc. 20th Ann. Conf.*, Princeton University, 301–306.

[19] Keerthi, S.S. and Gilbert, E.G. (1988) Optimal infinite-horizon feedback laws for a general class of constrained discrete-time systems: stability and moving-horizon approximations. *J. Optim. Theory Applics*, **57**, 265–293.

[20] Kleinman, D.L. (1970), An easy way to stabilize a linear constant system. *IEEE Trans. Automatic Control*, **15**, 692.

[21] Kwon, W.H., Bruckstein, A.M. and Kailath, T. (1983) Stabilizing state-feedback via the moving horizon method. *Int. J. Control*, **37**, 631–643.

[22] Kwon, W.H. and Pearson, A.E. (1977) A modified quadratic cost problem and feedback stabilization of a linear system. *IEEE Trans. Automatic Control*, **AC–22**, 838–842.

[23] Kwon, W.H. and Pearson, A.E. (1978) On feedback stabilization of time-varying discrete linear systems. *IEEE Trans. Automatic Control*, **AC–23**, 838–842.

[24] Leitmann, G. (1993) On one approach to the control of uncertain systems. *ASME J. Dynam. Syst., Measurem., and Control*, **115**, 373–380.

[25] Magaña, M.E. and Żak, S.H. (1988) Robust state feedback stabilization of discrete-time uncertain dynamical systems. *IEEE Trans. Aut. Control*, **AC–33**, 887–891.

[26] Mayne, D.Q. and Michalska, H. (1990) Receding horizon control of nonlinear systems. *IEEE Trans. Automatic Control*, **AC–35**, 814–824.

[27] Meadows, E.S. and Rawlings, J.B. (1993) Receding horizon control with an infinite horizon. *Proc. American Control Conference*, San Francisco, 2926–2930.

[28] Michalska, H. and Mayne, D.Q. (1991) Receding horizon control of nonlinear systems without differentiability of the optimal value function. *Systems and Control Letters*, **16**, 123–130.

[29] Michalska, H. and Mayne, D.Q. (1993) Robust receding horizon control of constrained nonlinear systems. *IEEE Trans. Automatic Control*, **AC–38**, 1623–1633 .

[30] Parisini, T. and Zoppoli, R. (1995) A receding-horizon regulator for nonlinear systems and a neural approximation. *Automatica*, **31**, 1443–1451.

[31] Polak E. and Yang T.H. (1993) Moving horizon control of linear systems with input saturation and plant uncertainty. 1, 2. *Int. J. Control*, **58**, 613–638, 639–663.

[32] Vassilaki, M., Hennet, J.C. and Bitsoris, G. (1988) Feedback control of linear discrete-time systems under state and control constraints. *Int. J. Control*, **47**, 1727–1735.

[33] Wredenhagen G.F. and Belanger P.R. (1994) Piecewise-linear LQ control for systems with input constraints. *Automatica*, **30**, 403–416.

7 ON GLOBAL CONTROLLABILITY OF TIME-INVARIANT NONHOMOGENEOUS BILINEAR SYSTEMS

R.R. MOHLER[1] and ALEX Y. KHAPALOV[2]

[1] *Department of Electrical and Computer Engineering, Oregon State University, Cornvallis, OR, USA*
[2] *Department of Pure and Applied Mathematics, Washington State University, Pulman, WA, USA*

1 Introduction

Bilinear systems evolve, as natural models or as accurate approximations to numerous dynamical processes in engineering, economics, biology, ecology, etc., and in other uses bilinear control may be implemented to improve controllability of an otherwise linear system [10–13]. In this note some sufficient conditions for complete controllability of the following nonhomogeneous BLS are discussed:

$$\dot{x} = Ax + \sum_{j=1}^{m}(N_j x + b_j)u_j, \quad x(0) = x_0, \quad t > 0. \tag{1.1}$$

Here A, N_j are $[n \times n]$-matrices, $b_j \in R^n$, $j = 1, \ldots, m$, and $\mathbf{u} = (u_1, \ldots, u_m)$ is an m-dimensional control. We further assume that $\mathbf{u} \in L^2(0, \infty; R^m)$.

There are a large number of publications relevant to this topic. Sufficient conditions for complete controllability of (1.1) were given in [12] assuming the existence of a pair of connected stable and unstable equilibrium points at which the vector field is nonsingular. Based on the "visualization" of the qualitative behavior of BLS on the plane, the planar case was completely resolved in [8]. Controllability of BLS has also been approached by differential geometric methods in terms of vector fields on manifolds in a series of papers [2, 1, 4, 3, 5] (see also the literature therein). Their results require the calculation of the Lie algebras of the corresponding polysystem and its accessible sets at every point of manifold. Verification of these conditions, on the other hand, is a difficult problem in the case of several dimensions, though in a number of specific situations they were detailed for the *homogeneous* BLS in [1, 4, 3]. This situation indicates that beyond the 2-D case the problem addressed in this paper is still rather open.

In recent work (see [7]) a new qualitative method to analyze the global complete controllability of nonhomogeneous BLS was proposed. It is based on a nonlinear decomposition technique which reduces the solution of the controllability problem for BLS to the analysis of the qualitative behavior of the following $2m+2$ *uncontrolled linear systems*:

$$\dot{x} = Ax, \quad t > 0, \tag{1.2}$$

$$\dot{x} = N_j x + b_j \text{ or } \dot{x} = -N_j x - b_j, \quad j = 1, \ldots, m, \quad t > 0. \tag{1.3}$$

More precisely, it was shown in [7] that one can embed the set of all continuous spline-trajectories starting from a given initial state and consisting of any finite combinations of the integral curves of (1.2), (1.3) (*their qualitative behavior is well known*) into the geometric closure (i.e., in R^n) of the set of the integral curves of (1.1) starting from the same point. The main goal of this note is to apply the general results of [7] to treat a number of specific situations in several space dimensions (i.e., $n > 2$).

The note is organized as follows. In Section 2 we introduce necessary preliminaries. In Section 3 several sufficient conditions for global complete controllability of (1.1) are discussed.

2 Preliminaries

A point $x_1 \in R^n$ is said to be *reachable from* a point $x_0 \in R^n$, if there exists a control $\mathbf{u}$ which in a finite time $T = T(x_0, x_1)$ drives system (1.1) from the state $x(0) = x_0$ to the state $x(T) = x_1$. System (1.1) is said to be *completely controllable in* R^n ($P \subset R^n$) over an infinite time horizon if any $x_1 \in R^n(P)$ is reachable from any $x_0 \in R^n(P)$. If P is an open neighborhood of a point $\bar{x}$, then we shall say that (1.1) is *locally controllable at* $\bar{x}$. We shall refer the set of all the points reachable from x_0 as *the reachable set from* x_0 and denote it by $\mathbf{R}(x_0)$.

We shall say that x_1 is reachable from x_0 due to (1.2)–(1.3) if there exists a continuous spline trajectory which connects x_0 with x_1 and consists of a finite number of pieces each of which is the integral curve of one of the uncontrolled systems (1.2)–(1.3). We shall refer to the set of all the points reachable from x_0 due to (1.2)–(1.3) as *the reachable set from* x_0 *due to* (1.2)–(1.3) and denote it by $\mathbf{R}_{un}(x_0)$. By $\mathbf{R}^-_{un}(x_0)$ we denote the reachable set from x_0 due to (1.3) and the system

$$\dot{x} = -Ax, \quad t > 0. \tag{2.1}$$

The following results were obtained in [7].

Theorem 2.1 *Let* $\exists \bar{x} \in \bigcap_{x_0 \in R^n} (\text{cl } \{\mathbf{R}_{un}(x_0)\} \bigcap \text{cl } \{\mathbf{R}^-_{un}(x_0)\})$ *at which (1.1) is locally controllable. Then (1.1) is completely controllable in* R^n. *(Here and elsewhere "cl" denotes the closure in* R^n.*)*

Theorem 2.2 *Let the pair* $\{A,\ B = \{b_1, \dots, b_m\}\}$ *be controllable, that is,*

$$\text{rank } \{B, AB, \dots, A^{n-1}B\} = n$$

and let the origin $\bar{0} \in \bigcap_{x_0 \in R^n} (\text{cl } \{\mathbf{R}_{un}(x_0)\} \bigcap \text{cl } \{\mathbf{R}^-_{un}(x_0)\})$. *Then (1.1) is completely controllable in* R^n.

The proof follows immediately from Theorem 2.1 by applying the well-known sufficient condition for local controllability of a nonlinear system at an equilibrium point, see, e.g., [9]. Theorems 2.1 and 2.2 yield the following general scheme.

Procedure 2.3 Verification of controllability of (1.1) in $\mathbf{R^n}$.

1. Find a point $\bar{x}$ of local controllability of (1.1).
2. Find the eigenvalues and eigenvectors of matrices $A, N_1, \dots, N_m$. This provides the information about $2m + 2$ phase-portraits of (1.2), (1.3), (2.1).
3. Check the reachability of an arbitrary neighborhood of $\bar{x}$ from any $x_0 \in R^n$ along the found phase-portraits associated accordingly with (1.2), (1.3) and (1.3), (2.1) by making use of suitable switches from one to another.

Remark 2.1 (i) Note that the existence of a point $\bar{x}$ required in Theorem 2.1 is a necessary condition for complete controllability of (1.1).

(ii) Sufficient conditions for complete controllability of (1.1) obtained in [12] and in [8] in the planar case are concerned with adequate diversity of the eigenvalues (*i.e.*, of phase-portraits) of the following m-parameter family of matrices

$$A + \sum_{j=1}^{m} N_j \mu_j$$

when $\mu_j,\ j = 1, \dots, m$ run over all R. However, these conditions are difficult to check in the general multidimensional case. Theorem 2.1 reduces the study of controllability of (1.1) to the study of the eigenvalues and eigenvectors of $m + 1$ matrices $A,\ N_j,\ j = 1, \dots, m$ only, though how far this reduction is sufficient for the complete solution of the problem is an open question yet.

3 Some Sufficient Conditions for Complete Controllability

In this section we apply Theorems 2.1 and 2.2 to investigate the problem of complete controllability of (1.1) in a number of specific situations.

Theorem 3.1 *Let there exist an index j^* such that $AN_{j^*}^{-1}b_{j^*} = 0$ and the pair*

$$\{A + N_{j^*}u_{j^*}^*,\ B = \{-N_1N_{j^*}^{-1}b_{j^*} + b_1, \ldots, -N_jN_{j^*}^{-1}b_{j^*} + b_j, \ldots, -N_mN_{j^*}^{-1}b_{j^*} + b_m\}_{j \neq j^*}\}$$

is controllable (see Theorem 2.2 in the above) for some constant $u_{j^*}^*$. Let all the eigenvalues of N_{j^*} have real parts of the same sign. Then (1.1) is completely controllable in R^n.

Proof By substitution $z = x + N_{j^*}^{-1}b_{j^*}$, system (1.1) is transformed into

$$\dot{z} = (A + N_{j^*}u_{j^*}^*)z + \sum_{j=1, j\neq j^*}^{m} (N_jz - N_jN_{j^*}^{-1}b_{j^*} + b_j\}u_j + N_{j^*}z(u_{j^*} - u_{j^*}^*), \qquad (3.1)$$

$$z(0) = z_0, \quad t > 0.$$

Obviously, if the latter system is completely controllable in R^n, then (1.1) is also completely controllable.

By the assumptions,

$$\text{rank } \{B, (A + N_{j^*}u_{j^*}^*)B, \ldots, (A + N_{j^*}u_{j^*}^*)^{n-1}B\} = n.$$

Hence, (3.1) is locally controllable at the origin.

Note next that for system (3.1) one of the corresponding equations (1.3) is as follows:

$$\dot{z} = \pm N_{j^*}z, \quad t > 0,$$

$$z(0) = z_0.$$

(To obtain this, one needs to replace control u_{j^*} by $\bar{u}_{j^*} = u_{j^*} - u_{j^*}^*$.) Due to the assumptions of Theorem 3.1, the origin is a unstable/stable (for $+/-$) equilibrium of the latter system. Therefore, the embedding

$$\bar{0} \in \bigcap_{z_0 \in R^n} (\text{cl } \{\mathbf{R}_{un}(z_0)\} \bigcap \text{cl } \{\mathbf{R}_{un}^-(z_0)\})$$

can be shown by employing its motions only. Applying Theorem 2.2 completes the proof of Theorem 3.1.

Theorem 3.2 *Let there exist a subset of indices j^*, $\{j_i\}_{i=1}^I$, $I \leq m-1$ such that $N_{j_i}, i = 1, \ldots I$ are null-matrices, $b_{j^*} = 0$, and the pair $\{N_{j^*},\ B = \{b_{j_1}, \ldots, b_{j_I}\}\}$ is controllable. Then system (1.1) is completely controllable in R^n.*

Proof *Step 1.* Under the assumptions of Theorem 3.2, we have

$$\dot{x} = (A + u_{j^*}N_{j^*})x + \sum_{j=1,\ j\neq j_i,\ i=1,\ldots,I}^{m} (N_jx + b_j)u_j + \sum_{i=1}^{I} b_{j_i}u_{j_i}, \quad x(0) = x_0, \quad t > 0.$$

This system will be locally controllable at the origin if

$$\text{rank } \{B, (A + u_{j^*}N_{j^*}B, \ldots, (A + u_{j^*}N_{j^*})^{n-1}B\} = n. \qquad (3.2)$$

By the assumptions, we have

$$\text{rank}\ \{B, N_{j^*}B, \ldots, N_{j^*}^{n-1}B\} = n,$$

which yields (3.2) for sufficiently large $|u_{j^*}|$. Hence, under the conditions of Theorem 3.2 system (1.1) is locally controllable at the origin.

Step 2. The argument of Theorem 2.1 given in [7] employs the idea that by increasing the magnitude of the control **u** the "drift" term Ax can be suppressed in (1.1), as though the time scales become different for the controlled and uncontrolled parts. This allows one to distinguish the set of "basic" motions (1.2), (1.3) that can be approximated in R^n with an arbitrary accuracy by actual integral curves of (1.1). By suitable scaling of the controls $u_{j^*}, u_{j_i}, i = 1, \ldots,$ while setting $u_j \equiv 0$ for $j \neq j^*, j_i, i = 1, \ldots, I$, it can be shown that the results of Theorems 2.1 and 2.2 hold true if (1.3) is complemented by the following system:

$$\dot{x} = N_{j^*}x + \sum_{i=1}^{I} b_{j_i}u_{j_i}, \quad t > 0.$$

Since it is completely controllable in R^n, the embedding

$$\bar{0} \in \bigcap_{x_0 \in R^n} (\text{cl}\ \{\mathbf{R}_{un}(x_0)\} \bigcap \text{cl}\ \{\mathbf{R}_{un}^{-}(x_0)\})$$

can be achieved by employing its motions. Applying Theorem 2.2 completes the proof. □

Theorem 3.3 **An iterative sufficient condition.** *Consider the following $(n+1)$-dimensional bilinear system:*

$$\begin{pmatrix} \dot{x} \\ \dot{z} \end{pmatrix} = \begin{pmatrix} a_{11} \ldots a_{1n} \ a_{1n+1} \\ \vdots \\ a_{n1} \ldots a_{nn} \ a_{nn+1} \\ 0 \ldots\ldots 0 \ a \end{pmatrix} \begin{pmatrix} x \\ z \end{pmatrix} + \tag{3.3}$$

$$\sum_{j=1}^{m} \left(\begin{pmatrix} n_{j11} \ldots n_{j1n} \ n_{j1n+1} \\ \vdots \\ n_{jn1} \ldots n_{jnn} \ n_{jnn+1} \\ 0 \ldots\ldots 0 \ n_j \end{pmatrix} \begin{pmatrix} x \\ z \end{pmatrix} + \begin{pmatrix} b_j \\ 0 \end{pmatrix} \right) u_j + \left(N_{m+1} \begin{pmatrix} x \\ z \end{pmatrix} + \mathbf{b}_{m+1} \right) u_{m+1},$$

$$A = \begin{pmatrix} a_{11} \ldots a_{1n} \\ \vdots \\ a_{n1} \ldots a_{nn} \end{pmatrix}, \quad N_j = \begin{pmatrix} n_{j11} \ldots n_{j1n} \\ \vdots \\ n_{jn1} \ldots n_{jnn} \end{pmatrix}, \tag{3.4}$$

where $x, b_j \in R^n$, $j = 1, \ldots, m$, $\mathbf{b}_{m+1} \in R^{n+1}$, $z \in R$. Let system (1.1) (corresponding to (3.4)) be completely controllable in R^n and (3.3) be locally controllable at the *origin. Then system (3.3) is completely controllable in* R^{n+1}, *if any point* $\{x, z\} \in R^{n+1}$, $z \neq 0$ *can be transferred due to system (3.3) into the subspace* $\{\{x, 0\} \mid x \in R^n\}$ *and any point* $\{x, z\} \in R^{n+1}$, $z \neq 0$ *can be reached along the system (3.3) from the aforementioned subspace.*

Proof We apply Theorem 2.2. To show that $\bar{0} \in \text{cl}(\mathbf{R}_{un}(\{x_0, z_0\}))\ \forall\ \{x_0, z_0\} \in R^{n+1}$, take any $\{x_0, z_0\}$ and, by the assumptions, transfer it first, say at some moment $t = t^*$, into the n-dimensional subspace $\{\{x, 0\} \mid x \in R^n\}$. We note next that, if one selects $u_{m+1}(t) \equiv 0$ for $t > t^*$, then (3.3) can be decomposed into two parts as follows:

$$\dot{x} = \begin{pmatrix} a_{11} \dots a_{1n}\ a_{1n+1} \\ \vdots \\ a_{n1} \dots a_{nn}\ a_{nn+1} \end{pmatrix} \begin{pmatrix} x \\ z \end{pmatrix} + \sum_{j=1}^{m} \left(\begin{pmatrix} n_{j11} \dots n_{j1n}\ n_{j1n+1} \\ \vdots \\ n_{jn1} \dots n_{jnn}\ n_{jnn+1} \end{pmatrix} \begin{pmatrix} x \\ z \end{pmatrix} + b_j \right) u_j, \quad (3.5a)$$

and

$$\dot{z} = \left(a + \sum_{j=1}^{m} n_j u_j \right) z. \quad (3.5b)$$

Since $z = 0$ is an equilibrium point for (3.5b), then for any $t > t^*$ one has $z(t) \equiv 0$, if $u_{m+1}(t) = 0$, *regardless* of the choice of controls $u_i, i = 1, \dots, m$. This implies that (3.5a) becomes independent of z:

$$\dot{x} = \begin{pmatrix} a_{11} \dots a_{1n} \\ \vdots \\ a_{n1} \dots a_{nn} \end{pmatrix} x + \sum_{j=1}^{m} \left(\begin{pmatrix} n_{j11} \dots n_{j1n} \\ \vdots \\ n_{jn1} \dots n_{jnn} \end{pmatrix} x + b_j \right) u_j.$$

The latter system is completely controllable in $\{\{x, 0\} \mid x \in R^n\}$, by the assumptions. Therefore, we can drive (3.3) from $\{x(t^*),\ z(t^*) = 0\}$ into an arbitrary neighborhood of the origin, while remaining in the subspace $\{\{x, 0\} \mid x \in R^n\}$. A similar argument applies to show that $\bar{0} \in \text{cl}(\mathbf{R}^-_{un}(\{x_0, z_0\}))$, $\forall \{x_0, z_0\} \in R^{n+1}$. This completes the proof of Theorem 3.3.

Corollary 3.4 Assume in addition to the conditions of Theorem 3.3 that $a \neq 0$ and that all the eigenvalues of the matrix A have real parts of the same sign as a. Then system (3.3) is completely controllable in R^{n+1}, if:

(i) in the case $a > 0$ any point $\{x, z\} \in R^{n+1}$, $z \neq 0$ can be transferred due to system (3.3) into the subspace $\{\{x, 0\} \mid x \in R^n\}$;
(ii) in the case $a < 0$ any point $\{x, z\} \in R^{n+1}$, $z \neq 0$ can be reached due to system (3.3) from the subspace $\{\{x, 0\} \mid x \in R^n\}$.

The proof follows by applying Theorem 3.3, noticing that the origin is an unstable/stable node of (3.3) in accordingly the cases i/ii.

Remark 3.1 Assumption (i) in Corollary 3.4 can be achieved if for any point $\{x_0, z_0\}$ there exists a spline curve which starts from it, is composed from the integral curves of (1.2), (1.3), and which *crosses* the subspace $\{\{x, 0\} \mid x \in R^n\}$ (this is necessary here, because systems (1.3) are *only* "approximations" of some actual motions of (1.1)). To satisfy (ii), one can verify this property for (2.1), (1.3).

Example Consider the following 3-*D* system:

$$\begin{aligned}\dot{x}_1 &= x_1 - 2x_2 + (x_1 + 1)u_1 + x_1 u_2 \\ \dot{x}_2 &= x_1 + x_2 + 2x_2 u_2 \\ \dot{x}_3 &= x_3 + x_3 u_1 + (3x_3 - 1)u_2\end{aligned} \tag{3.6}$$

Step 1. In this example the pair $\{A, \{b_1, b_2\}\}$, where

$$A = \begin{pmatrix} 1 & -2 & 0 \\ 1 & 1 & 0 \\ 0 & 0 & 1 \end{pmatrix}, \quad N_1 = \begin{pmatrix} 1 & 0 & 0 \\ 0 & 0 & 0 \\ 0 & 0 & 1 \end{pmatrix}, \quad N_2 = \begin{pmatrix} 1 & 0 & 0 \\ 0 & 2 & 0 \\ 0 & 0 & 3 \end{pmatrix},$$

$$b_1 = \begin{pmatrix} 1 \\ 0 \\ 0 \end{pmatrix}, \quad b_2 = \begin{pmatrix} 0 \\ 0 \\ -1 \end{pmatrix},$$

is controllable, by the rank condition. Hence, system (3.6) is locally controllable at the origin. Now we intend to apply Corollary 3.4(i) with $a = 1$.

Step 2. The uncontrolled drift system has the eigenvalues

$$\lambda_{1,2} = 1 \pm i\sqrt{2}, \ \lambda_3 = 1.$$

Hence, the origin is the unstable equilibrium point for (3.6), from which one can conclude that $\bar{0} \in \mathrm{cl}\{\mathbf{R}^-_{un}(x_0)\} \ \forall x_0 \in R^3$.

Step 3. The orthogonal projection of the controlled part on the plane $x_1 \bar{0} x_2$ for $u_2 = 0$ is the following:

$$\begin{aligned}\dot{x}_1 &= x_1 - 2x_2 + (x_1 + 1)u_1, \\ \dot{x}_2 &= x_1 + x_2.\end{aligned} \tag{3.7}$$

To show complete controllability in R^2, we shall apply Theorem 2.2 (which is easier to do in this planar case). Note that the corresponding pair $\{A, b\}$, where

$$A = \begin{pmatrix} 1 & -2 \\ 1 & 1 \end{pmatrix}, \quad b = \begin{pmatrix} 1 \\ 0 \end{pmatrix}$$

is controllable. Hence, 2-*D* system (3.7) is locally controllable in some neighborhood of the origin in R^2. The drift system associated with it,

$$\dot{x}_1 = x_1 - 2x_2 \tag{3.8}$$

$$\dot{x}_2 = x_1 + x_2$$

has the eigenvalues $\lambda_{1,2} = 1 \pm i\sqrt{2}$, which gives the unstable focus at the origin. Hence, in view of (2.1), $\bar{0} \in \mathrm{cl}\{\mathbf{R}^-_{un}(x^*_0)\}\ \forall x^*_0 = \{x_{10}, x_{20}\} \in R^2$. To show that $\bar{0}$ is reachable due to the corresponding systems (1.2), (1.3) from any point $x^*_0 \in R^2$, it is sufficient first to transfer x^*_0 due to the drift system (3.8) on the axis $\bar{0}x_1$ at any point $\{x^{**}_1, 0\}, x^{**}_1 > -1$ and then to drive it to the origin (or into its controllable neighborhood) along this axis due to the corresponding systems in (1.3), that is,

$$\dot{x}_1 = \pm(x_1 + 1) \tag{3.9}$$

$$\dot{x}_2 = 0$$

This is possible, because $x_1 = -1$ is the unstable/stable node of the first equation in (3.9) (for $+/-$). This provides the required complete controllability of 2-*D* system (3.7) by Theorem 2.2.

Step 4. Let us show now that (3.6) can be transferred from any point $x \in R^3$ on the plane $x_1\bar{0}x_2$. Indeed, its controlled part generates the following two systems (see (1.3)):

$$\dot{x}_1 = \pm(x_1 + 1), \quad \dot{x}_2 = 0, \quad \dot{x}_3 = \pm x_3; \tag{3.10}$$

$$\dot{x}_1 = \pm x_1, \quad \dot{x}_2 = \pm 2x_2, \quad \dot{x}_3 = \pm(3x_3 - 1). \tag{3.11}$$

Motions (3.10) can transfer any point of R^3 into the semispace $\{x \in R^3 \mid x_3 \leq \frac{1}{3}\}$. Since the point $(0, 0, 1/3)$ is the unstable/stable node of (3.11) (depending on the sign), in view of Remark 3.1 this then allows one to transfer system (3.6) to the plane $x_1\bar{0}x_2$. Combining Steps 1–4 with Corollary 3.4(i) yields the complete controllability of (3.6) in R^3.

References

[1] Bonnard, B. (1981) Contrôlabilité des systémes bilinéaires. *Math. Systems Theory*, **15**, 79–92.

[2] Boothby, W.M. and Wilson, E.N. (1979) Determination of the transitivity of bilinear systems. *SIAM J. Contr. Optim.*, **17**(2), 212–221.

[3] Gauthier, J.P. and Bornard, G. (1982) Contrôlabilité des systémes bilinéares. *SIAM J. Contr. Optim.*, **20**(3), 377–384.

[4] Jurdjevic, V. and Kupka, I. (1981) Control systems on semi-simple Lie groups and their homogeneous spaces. *Ann. Inst. Fourier, Grenoble*, **31**(4), 151–179.

[5] Jurdjevic, V. and Sallet, G. (1984) Controllability properties of affine fields. *SIAM J. Contr. Optim.*, **22**(3), 501–508.

[6] Jurdjevic, V. and Sussmann, H. (1972) Control systems on Lie groups. *J. Diff. Equations*, **12**, 313–329.

[7] Khapalov, A.Y. and Mohler, R.R. (1996) Reachable sets and controllability of bilinear time-invariant systems: A qualitative approach. *IEEE Trans. on Autom. Contr.*, **41**, 1340–1346.

[8] Koditschek, D.E. and Narendra, K.S. (1985) The controllability of planar bilinear systems. *IEEE Trans. Autom. Control*, **AC-30**, 87–89.
[9] Marcus, L. (1966) Controllability of nonlinear processes. *SIAM J. Control,* **3**, 78–90.
[10] Mohler, R.R. (1973) *Bilinear Control Process*. Academic Presses, New York.
[11] Mohler, R.R. (1991) *Nonlinear Systems: V.2 Application to Bilinear Control*. Prentice-Hall, Englewood Cliffs, NJ.
[12] Rink, R.E. and Mohler, R.R. (1968) Completely controllable bilinear systems. *SIAM J. Control*, **6**, 477–486.
[13] Rink, R.E. and Mohler, R.R. (1971) Reachable zones for equicontinuous bilinear control process, *Int. J. Control*, **14**, 331–339.

8 ROBUST SUBOPTIMAL CONTROL OF CONSTRAINED PARABOLIC SYSTEMS UNDER UNCERTAINTY CONDITIONS*

BORIS S. MORDUKHOVICH and KAIXIA ZHANG

Department of Mathematics, Wayne State University, Detroit, MI 48202, USA

1 Introduction

This paper is concerned with robust control design of constrained parabolic systems under uncertain disturbances (perturbations) and feedback controllers in the mixed boundary conditions. Our interest in such problems was originally motivated by applications to automatic control of groundwater regimes in irrigation networks where the objective was to neutralize negative effects of uncertain weather conditions; see [9]. Here we consider a more general class of parabolic control systems that have a broad spectrum of practical applications.

Dynamical processes in such systems are described by linear second-order parabolic equations with boundary controllers and pointwise state and control constraints. One of the most remarkable features of these processes is their functioning in the presence of uncertain perturbations when only an admissible region is given and no probabilistic information is available. A natural approach to control design of uncertain systems is *minimax synthesis* (principle of guaranteed result) which provides the best system performance under worst perturbations and ensures acceptable (at least stable) behavior under any admissible perturbations.

Such a minimax approach to feedback control design is related to theories of differential games and robust H_∞-control; see [2, 5, 6] and their references. However, we are not familiar with any results in these theories that could be directly applied to the parabolic systems considered below under *hard* (pointwise) control and state constraints.

* This research was partly supported by the National Science Foundation under grants DMS-9404128 and DMS-9704751 and by the NATO contract CRG-950360.

In this paper we developed an effective multi-step approximation procedure to design *suboptimal* feedback controllers for constrained parabolic systems. This procedure is initiated in [9, 11] for the case of one-dimensional heat-diffusion equations and takes into account certain specific features of the parabolic dynamics with infinite horizon. Related results for more general parabolic equations with both Dirichlet and mixed boundary conditions are presented in [11–13].

This paper contains new results for the case of mixed boundary controllers. The results obtained include a justification of a suboptimal three-positional control structure with subsequent optimization of its parameters. The main goal is to ensure the desired state performance within required state constraints for all admissible perturbations and to minimize the given (energy type) cost functional in the case of maximal ones. Moreover, we obtain effective stability conditions to exclude unacceptable self-vibrating regimes for nonlinear closed-loop control systems with the given parabolic dynamics and three-positional mixed boundary controllers.

The paper is organized as follows. In Section 2 we formulate the robust feedback control problem of our study and present the main properties of the parabolic dynamics used in the sequel. Section 3 is devoted to solving first-order ODE approximation problems under maximal perturbations that allows us to justify a suboptimal structure of boundary controls in the parabolic system. In Section 4 we optimize parameters of this structure along the parabolic dynamics. Section 5 deals with computing a feedback boundary controller that ensures the best system behavior under maximal perturbations and keeps transients within the required state constraint region for any admissible disturbances on a sufficiently large control interval. The concluding Section 6 provides stability conditions for the class of nonlinear closed-loop control systems under consideration.

2 Problem Formulation and Basic Representations

Let $\Omega \subset \mathbf{R}^n$ be a bounded open set with the closure $\mathrm{cl}\Omega$. Assume that the boundary Γ of Ω is a C^∞-manifold of dimension $n-1$ and that locally Ω lies on one side of Γ. Let a_0 and $a_{ij}, i,j = 1,2,\ldots,n$, be given real-valued functions with the properties a_0, $a_{ij} \in C^\infty(\mathrm{cl}\Omega)$,

$$a_{ij}(x) = a_{ji}(x) \forall i,j = 1,\ldots,n, \quad x \in \Omega;$$

$$\sum_{i,j=1}^{n} a_{ij}(x)\xi_i\xi_j \geq \beta \sum_{i=1}^{n} \xi_i^2, \quad \beta > 0 \ \forall \xi \in \mathbf{R}^n, \quad x \in \Omega. \tag{2.1}$$

Observe that the linear operator

$$A := -\sum_{i,j=1}^{n} \frac{\partial}{\partial x_i}\left(a_{ij}(x)\frac{\partial}{\partial x_j}\right) + a_0(x) \tag{2.2}$$

is self-adjoint and *uniformly strongly elliptic* on $L^2(\Omega)$ due to (2.1).

In this paper we study the following parabolic system with the mixed boundary conditions:

$$\begin{cases} \frac{\partial y}{\partial t} + Ay = w(t) & \text{a.e. in } Q := (0,T) \times \Omega \\ y(0,x) = 0, \quad x \in \Omega \\ \left(\alpha y + \frac{\partial y}{\partial \nu_A}\right)|_\Sigma = u(t), & \Sigma := (0,T) \times \Omega \end{cases} \tag{2.3}$$

where $\alpha > 0$ and the normal derivative $\dfrac{\partial}{\partial \nu_A}$ is defined by

$$\frac{\partial}{\partial \nu_A} := \sum_{i,j=1}^{n} a_{ij}(x) \frac{\partial}{\partial x_j} \cos(\mathbf{n}, x_i).$$

In what follows we treat $w(\cdot)$ as an *uncertain disturbance* perturbing the system, and $u(\cdot)$ as a *control* that can be chosen to achieve a required system performance. It is well known that for each $(u,w) \in L^2(0,T) \times L^2(0,T)$ system (2.3) has a unique *generalized solution* in the sense of [7]. Moreover, it follows from [8] that this solution $y = y(t,x)$ is *continuous* on $\text{cl}Q := [0,T] \times \text{cl}\Omega$.

Given positive numbers $\bar{a}$, $\underline{a}$, $\bar{b}$, and $\underline{b}$, we define the sets of *admissible controls* $u(\cdot)$ and admissible *uncertain perturbations* $w(\cdot)$ by, respectively,

$$U_{ad} := \{u \in L^2(0,T) \mid u(t) \in [-\bar{a}, \underline{a}] \quad \text{a.e.} \quad t \in [0,T]\},$$
$$W_{ad} := \{w \in L^2(0,T) \mid w(t) \in [-\underline{b}, \bar{b}] \quad \text{a.e.} \quad t \in [0,T]\}.$$

Suppose that $x_0 \in \Omega$ is a given point at which one measures the system performance and that $\eta > 0$ is an assigned constant. We consider the following *minimax feedback control problem*:

(P) minimize

$$J(u) = \max_{w(\cdot) \in W_{ad}} \int_0^T |u(y(t,x_0))| dt$$

over $u(\cdot) \in U_{ad}$ subject to (2.3) with the *pointwise state constraints*

$$|y(t,x_0)| \le \eta \quad \forall t \in [0,T] \tag{2.4}$$

and the *feedback control law* formed by

$$u(t) = u(y(t,x_0)) \tag{2.5}$$

through the *mixed boundary conditions* in (2.3).

We always assume that there exists at least one triplet $(u,w,y) \in U_{ad} \times W_{ad} \times C(\text{cl}Q)$ such that it is *feasible* for problem (P), i.e., satisfies all the constraints.

Note that we do not have any available information about uncertain perturbations $w(t)$ except the given bounds $\{-\underline{b},\ \bar{b}\}$ of their admissible values. The objective in (P) is to find a feedback control function $u = u(y) \in [-\bar{a}, \underline{a}]$ of the intermediate state $y = y(t,x_0)$ that keeps the system performance within the constraint region (2.4) for all admissible perturbations and minimizes the given cost functional in the case of worst perturbations. This is a *minimax robust control* problem for uncertain distributed parameter systems under hard state and control constraints. Problems of this kind are among the most difficult ones in control theory, and we are not familiar with any effective methods to solve such problems in full generality. Let us describe an approach to solving (P) that takes into account certain specific features of parabolic systems and allows us to compute a feasible *suboptimal* (in some sense) feedback control.

Our approach employs the Fourier series *spectral representation* of solutions of the parabolic system (2.3). To this end we consider the *eigenvalue problem*

$$\begin{cases} -A\varphi + \lambda\varphi = 0 \\ \left.\left(\alpha\varphi + \dfrac{\partial\varphi}{\partial\nu_A}\right)\right|_{\Sigma} = 0 \end{cases} \tag{2.6}$$

involving eigenvalues λ and eigenfunctions φ. It is well known (see, e.g., [1]) that under the general assumptions made there exists a sequence of solutions $\{\lambda_k, \varphi_k\}_{k\in\mathbf{N}}$ to (2.6) such that

$$\{\varphi_k\}_{k\in\mathbf{N}} \text{ is a complete orthonormal basis in } L^2(\Omega) \text{ and}$$
$$\lambda_k = ck^{\frac{2}{n}} + o(k^{\frac{2}{n}}) \text{ for some } c > 0.$$

Consider the numbers

$$\mu_k := \int_\Omega \varphi_k(x)dx \quad \text{and} \quad \nu_k := \int_\Gamma \varphi_k(\zeta)d\sigma_\zeta$$

where $d\sigma_\zeta$ denotes the surface measure. The following result [7, 8] provides the basic spectral representation of solutions of the parabolic system (2.3).

Proposition 1 *Let $(u, w) \in L^2(0, T) \times L^2(0, T)$. Then the corresponding solution $y(t, x)$ of system* (2.3) *is continuous on* cl$Q$ *and is represented in the form*

$$y(t,x) = \sum_{k=1}^{\infty}\left(\mu_k \int_0^t w(\theta)e^{\lambda_k\theta}d\theta + \nu_k \int_0^t u(\theta)e^{\lambda_k\theta}d\theta\right)e^{-\lambda_k t}\varphi_k(x) \tag{2.7}$$

where the series converges strongly in $L^2(\Omega)$ for each $t \in [0, T]$.

Employing the *maximum principle* for parabolic equations (cf. [4]), one gets *monotonicity properties* of transients in (2.3) with respect to both controls and perturbations that play a crucial role in what follows.

Proposition 2 *Let $(u_i, w_i) \in L^2(0, T) \times L^2(0, T)$ and let $y_i(\cdot)$ be the corresponding generalized solution to* (2.3) *for $i = 1,$. Then*

$$y_1(t,x) \geq y_2(t,x) \qquad \forall (t,x) \in Q$$

if $u_1(t) \geq u_2(t)$ and $w_1(t) \geq w_2(t)$ for all $t \in [0, T]$.

Remember that the control objective is to keep transients within the given state constraints under any admissible perturbations. Then Proposition 2 infers that *the bigger magnitude of a perturbation is, the more control of the opposite sign should be applied* to neutralize the perturbation and ensure the required state performance. This makes us consider feedback control laws with the *compensation property*

$$u(y) \le u(\tilde{y}) \quad \text{if} \quad y \ge \tilde{y} \quad \text{and} \quad y \cdot u(y) \le 0 \qquad \forall y, \tilde{y} \in \mathbf{R}. \tag{2.8}$$

The latter property implies that

$$\int_0^T |u(y(t))|\, dt \ge \int_0^T |u(\tilde{y}(t))|\, dt \ \text{ if } \ y(t) \ge \tilde{y}(t) \ge 0 \ \text{ or } \ y(t) \le \tilde{y}(t) \le 0 \ \forall t \in [0, T],$$

i.e., the compensation of bigger (by magnitude) perturbations requires more cost with respect to the maximized cost functional in (P). This allows us to seek a *suboptimal control structure* in (P) by examining the control response to feasible perturbations of the *maximal magnitudes* $w(t) = \bar{b}$ and $w(t) = -\underline{b}$ for all $t \in [0, T]$.

3 Approximation Problems

In this section we develop multi-step approximation procedures to justify an acceptable structure of feasible suboptimal controls for problem (P).

Let $u = u(y)$ be a given feedback control law in (P). Then for any given perturbation $w = w(t)$ we have an open-loop control realization $u(t) = u(y(t, x_0))$ due to system (2.3). We consider only *feasible* pairs $(u, w) \in U_{ad} \times W_{ad}$ such that the corresponding transient $y(t, x_0)$ satisfies the state constraints (2.4). For any natural number $N = 1, 2, \ldots$ we denote

$$y^N(t, x) := \sum_{k=1}^{N} \left(\mu_k \int_0^t w(\theta) e^{\lambda_k \theta} d\theta + \nu_k \int_0^t u(\theta) e^{\lambda_k \theta} d\theta \right) e^{-\lambda_k t} \varphi_k(x)$$

and conclude that for all $t \in [0, T]$

$$y^N(t, \cdot) \to y(t, \cdot) \text{ strongly in } L^2(\Omega) \text{ as } N \to \infty$$

due to Proposition 1. Moreover, considering $y^N(t, x)$ at the point of observation $x = x_0$, we get $y^N(t, x_0) = \sum_{k=1}^{N} y_k(t)$ with

$$\begin{cases} \dfrac{dy_k}{dt} = \lambda_k y_k + \mu_k w + \nu_k u \\ y_k(0) = 0, \quad k = 1, 2, \ldots, N. \end{cases}$$

Thus Proposition 1 allows us to approximate the original parabolic system (2.3) by systems of ordinary differential equations.

In what follows we assume that the eigenvalues in (2.6) satisfy the conditions

$$0 < \lambda_1 < \lambda_k, \quad k = 2, 3, \dots \tag{3.1}$$

that always hold, e.g., when $A = \Delta$ is the Laplacian. One can observe that under (3.1) the first term *asymptotically dominates* in the series (2.7) as $t \to \infty$. On this basis, we examine the case of $N = 1$ in the above ODE system to justify a *suboptimal* control structure for the original problem.

Taking into account the discussion after Proposition 2 as well as the *symmetry* of (P) relative to the origin, we consider the following *open-loop optimal control problem* with the admissible control set

$$\bar{U}_{ad} := \{u(\cdot) \in U_{ad} \mid -\bar{a} \le u(t) \le 0 \text{ a.e. } t \in [0, T]\}$$

in response to the upper level maximal perturbation $w(t) = \bar{b}$ on $[0, T]$:

$(\bar{\mathrm{P}}_1)$ minimize $\bar{J}(u) = -\int_0^T u(t)dt$

over $u(\cdot) \in \bar{U}_{ad}$ subject to

$$\begin{cases} \dfrac{dy}{dt} = -\lambda_1 y + \varphi_1(x_0)(\mu_1 \bar{b} + \nu_1 u(t)) & \text{a.e.} \quad t \in [0, T] \\ y(0) = 0 \end{cases} \tag{3.2}$$

and the state constraint

$$y(t) \le \eta \qquad \forall t \in [0, T]. \tag{3.3}$$

The symmetric case of $w(t) = -\underline{b}$ at the lower boundary level can be considered similarly and actually can be reduced to $(\bar{\mathrm{P}}_1)$.

Note that the presence of state constraints relates $(\bar{\mathrm{P}}_1)$ to the class of the most complicated optimal control problems for ODE systems. It is well known that in general their solutions involve *Borel measures* that make them fairly difficult for applications; see [3]. We avoid such difficulties by developing an *approximation procedure* in the vein of [10] to replace $(\bar{\mathrm{P}}_1)$ by a parametric family of standard optimal control problems with *no* state constraints. To solve approximation problems we employ the *Pontryagin maximum principle* [15] that provides *necessary and sufficient* conditions for optimality of approximating solutions. It happens that optimal controls to approximation problems contain both *bang-bang and singular modes*. Passing to the limit, we obtain in this way an *exact solution* to the state-constrained problem $(\bar{\mathrm{P}}_1)$ that *does not involve any measure*. The results obtained show that the state constraint (3.3) in $(\bar{\mathrm{P}}_1)$ turns out to be a *regularization factor*. Such a surprising conclusion is due to the specifics of problems like $(\bar{\mathrm{P}}_1)$ reflecting the parabolic dynamics. The reader can find more details in [11, 14].

Theorem 3 *Let* $\mu_1\varphi_1(x_0)\bar{b} > \lambda_1\eta$. *Assume in addition that either*

$$\tau_1 := \frac{1}{\lambda_1}\ln\frac{\mu_1\varphi_1(x_0)\bar{b}}{\mu_1\varphi_1(x_0)\bar{b} - \lambda_1\eta} \geq T \tag{3.4}$$

or

$$\mu_1\varphi_1(x_0)\bar{b} - \bar{a}\nu_1\varphi_1(x_0) \leq \lambda_1\eta.$$

Then system (3.2), (3.3) *is controllable, i.e., there is* $u(\cdot) \in \bar{U}_{ad}$ *such that the corresponding trajectory of* (3.2) *satisfies the state constraint* (3.3). *Moreover, problem* ($\bar{\text{P}}_1$) *admits an optimal control of the form*

$$\bar{u}_1(t) = \begin{cases} 0 & \text{if } t \in [0, \bar{\tau}_1] \\ \dfrac{\lambda_1\eta - \mu_1\varphi_1(x_0)\bar{b}}{\nu_1\varphi_1(x_0)} & \text{if } t \in [\bar{\tau}_1, T] \end{cases} \tag{3.5}$$

where $\bar{\tau}_1 = \min\{\tau_1, T\}$ *with* τ_1 *given by* (3.4).

Note that in (3.5) we have only one switching from the original bang-bang level to an *intermediate singular mode*. For the symmetric problem ($\underline{\text{P}}_1$) in which the system is operated under the lower level maximal perturbation $w(t) = -\underline{b}$, one can obtain the corresponding results from Theorem 3 by changing signs of the state and control variables.

According to the above discussions, the optimal controls derived for problems ($\bar{\text{P}}_1$) and ($\underline{\text{P}}_1$) can be viewed as *first-order suboptimal solutions* to the open-loop control problems arising from the original problem (P) under the maximal perturbations $w(t) = \bar{b}$ and $w(t) = -\underline{b}$. In the next section we show this simple structure to be justified as a *suboptimal control structure* for the original problem under maximal perturbations and then optimize its parameters along the *parabolic dynamics over a large control interval.*

4 Optimal Control Under Maximal Perturbations

Let us consider the following optimal control problem for the original parabolic system (2.3) under the upper level maximal perturbation $w(t) = \bar{b}$ on $[0, T]$:

($\bar{\text{P}}$) minimize $\bar{J}(u) = -\int_0^T u(t)dt$
subject to system (2.3), state constraint (3.3), and boundary controls $u(\cdot) \in \bar{U}_{ad}$ of the form

$$u(t) = \begin{cases} 0 & \text{for } 0 \leq t < \tau \\ -\bar{u} & \text{for } \tau \leq t \leq T. \end{cases} \tag{4.1}$$

We choose the mixed boundary control structure (4.1) according to the results in Section 3 that justify its suboptimality under maximal perturbations. To solve ($\bar{\text{P}}$)

one should find optimal parameters $\bar{u} \in [0, \bar{a}]$ and $\tau \in [0, T]$ in (4.1) which enforce the state constraint (3.3) along the parabolic dynamics (2.3) and minimize the given cost functional. In what follows we suppose that the control interval $[0, T]$ is *sufficiently large* and examine the *asymptotics* of optimal solutions as $T \to \infty$ based on assumption (3.1). It turns out that under this assumption optimal processes in $(\bar{\text{P}})$ possess a kind of *turnpike property* that simplifies the solution while passing to the infinite horizon.

To formulate the main results we need to introduce the following numbers

$$\gamma := \sum_{k=1}^{\infty} \frac{\mu_k \varphi_k(x_0)}{\lambda_k} \quad \text{and} \quad \rho := \sum_{k=1}^{\infty} \frac{\nu_k \varphi_k(x_0)}{\lambda_k}$$

that are positive under the assumptions made in the next theorem.

Theorem 4 *In addition to the basic assumptions let us suppose that $\nu_1 > 0$ and*

$$0 < \gamma \bar{b} - \eta < \min\left\{ \bar{a}\rho,\ \frac{\rho \mu_1 \bar{b}}{\nu_1} \right\}.$$

Consider the transcendental equation

$$\sum_{k=1}^{\infty} \frac{\varphi_k(x_0)}{\lambda_k} e^{-\lambda_k T} [\nu_k(\gamma \bar{b} - \eta) e^{\lambda_k \tau} - \rho \mu_k \bar{b}] = 0 \tag{4.2}$$

which has a unique solution $\tau = \bar{\tau}(T) \in (0, T)$ for all T sufficiently large. Then any control (4.1) with

$$\bar{u} = \frac{\gamma \bar{b} - \eta}{\rho} \tag{4.3}$$

is feasible in $(\bar{\text{P}})$ for all positive $\tau \leq \bar{\tau}(T)$, being optimal for this problem when $\tau = \bar{\tau}(T)$. Moreover, $\bar{\tau}(T) \downarrow \bar{\tau}$ as $T \to \infty$ where the asymptotically optimal switching time $\bar{\tau}$ is computed by

$$\bar{\tau} = \frac{1}{\lambda_1} \ln \frac{\rho \mu_1 \bar{b}}{\nu_1 (\gamma \bar{b} - \eta)}. \tag{4.4}$$

The proof of this theorem follows the scheme of [11] for the case of Dirichlet boundary conditions. Let us observe that control (4.1) with parameters (4.3) and (4.4) is feasible for problem $(\bar{\text{P}})$ on the interval $[0, T]$ with an arbitrarily large T. Moreover, $\bar{\tau}$ is the *maximal* one among all switching times in (4.1) that keeps the state constraint (3.3) on the whole infinite interval $[0, \infty)$. Therefore, this *asymptotically optimal control* with the infinite horizon is *suboptimal* for the given problem $(\bar{\text{P}})$ on $[0, T]$ where T is sufficiently large.

Similar results hold for the symmetric optimal control problem $(\underline{\text{P}})$ corresponding to the lower level maximal perturbation. This problem consists of minimizing the cost functional $\underline{J}(u) = \int_0^T u(t)dt$ subject to the parabolic system (2.3) with $w(t) = -\underline{b}$ on $[0, T]$, admissible boundary controls $0 \leq u(t) \leq \underline{a}$ of the form

$$u(t) = \begin{cases} 0 & \text{for } 0 \leq t < \tau \\ \underline{u} & \text{for } \tau \leq t \leq T, \end{cases} \tag{4.5}$$

and the state constraint $-\eta \leq y(t)$ for all $t \in [0, T]$.

Utilizing symmetric arguments, we justify the optimality of control (4.5) with

$$\underline{u} = \frac{\gamma \underline{b} - \eta}{\rho} \tag{4.6}$$

and the switching time $\tau = \underline{\tau}(T)$ satisfying (4.2) for $-\underline{b}$. One has $\underline{\tau}(T) \downarrow \underline{\tau}$ as $T \to \infty$ where the asymptotically optimal switching time is computed by

$$\underline{\tau} = \frac{1}{\lambda_1} \ln \frac{\rho \mu_1 \underline{b}}{\nu_1(\gamma \underline{b} - \eta)}.$$

5 Feedback Control Design for the Parabolic System

Let us return to the original feedback control problem (P) and assume in what follows that its initial data satisfy all the assumptions in Theorem 4 as well as the symmetric ones for the lower level maximal perturbation $w(t) = -\underline{b}$. Based on the results above, we consider the following *three-positional feedback control law* in (2.5):

$$u(y) = \begin{cases} -\bar{u} & \text{if } y \geq \bar{\sigma} \\ 0 & \text{if } -\underline{\sigma} < y < \bar{\sigma} \\ \underline{u} & \text{if } y \leq -\underline{\sigma} \end{cases} \tag{5.1}$$

that obviously satisfies the compensation property (2.8). We have established that structure (5.1) is *suboptimal* (optimal to the first order) with respect to the objective in (P) under the realization of the maximal boundary perturbations $w(\cdot) = \bar{b}$ and $w(\cdot) = -\underline{b}$. Furthermore, we computed optimal control parameters corresponding to the maximal perturbations with their asymptotics on the infinite horizon. Now our goal is to determine optimal parameters of the feedback control law (5.1) ensuring the desired behavior of the closed-loop system (2.3), (2.5), (5.1).

Let $\bar{u}$ and $\underline{u}$ in (5.1) be computed by formulas (4.3) and (4.6), respectively. Under the assumptions made one obviously has $u(\cdot) \in U_{ad}$ for any control realization $u(t) = u(y(t, x_0))$ corresponding to an arbitrary $w(\cdot) \in W_{ad}$. Moreover, these control values ensure the transient *stabilization* as $t \to \infty$ within the required state interval $[-\eta, \eta]$ for any admissible perturbations. However, the state constraints (2.4) may be violated for some $t \in [0, T]$ if the *dead region* $[-\underline{\sigma}, \bar{\sigma}]$ is not properly designed. The next theorem determines optimal values of $\underline{\sigma}$ and $\bar{\sigma}$ such that the closed-loop control system exhibits the best possible behavior under the maximal perturbations and keeps transients within the given state constraints for any admissible perturbations on a large control interval $[0, T]$. The proof is based on the transient monotonicity with respect to both controls and perturbations; cf. [11, 14].

Theorem 5 *Under the assumptions made we consider the feedback control (5.1) with $\bar{u}$ and $\underline{u}$ computed in (4.3) and (4.6), respectively. Let*

$$\bar{\sigma}(T) := \bar{b}\left(\gamma - \sum_{k=1}^{\infty} \frac{\mu_k \varphi_k(x_0)}{\lambda_k} e^{-\lambda_k \bar{\tau}(T)}\right),$$

$$\underline{\sigma}(T) := \underline{b}\left(\gamma - \sum_{k=1}^{\infty} \frac{\mu_k \varphi_k(x_0)}{\lambda_k} e^{-\lambda_k \underline{\tau}(T)}\right),$$

where $\bar{\tau}(T)$ and $\underline{\tau}(T)$ are the corresponding unique solutions to (4.2) *and its counterpart for $-\underline{b}$. Then the control law* (5.1) *is feasible for any perturbations $w(\cdot) \in W_{ad}$ and optimal in the case of maximal perturbations when T is sufficiently large. Moreover, $\bar{\sigma}(T) \downarrow \bar{\sigma}$ and $\underline{\sigma}(T) \downarrow \underline{\sigma}$ as $T \to \infty$ where the positive numbers*

$$\bar{\sigma} := \bar{b}\left(\gamma - \sum_{k=1}^{\infty} \frac{\mu_k \varphi_k(x_0)}{\lambda_k} \left[\frac{\nu_1(\gamma \bar{b} - \eta)}{\rho \mu_1 \bar{b}}\right]^{\frac{\lambda_k}{\lambda_1}}\right), \tag{5.2}$$

$$\underline{\sigma} := \underline{b}\left(\gamma - \sum_{k=1}^{\infty} \frac{\mu_k \varphi_k(x_0)}{\lambda_k} \left[\frac{\nu_1(\gamma \underline{b} - \eta)}{\rho \mu_1 \underline{b}}\right]^{\frac{\lambda_k}{\lambda_1}}\right) \tag{5.3}$$

form the maximal dead region $[-\underline{\sigma}, \bar{\sigma}]$ under which feedback (5.1) *keeps the state constraints* (2.4) *on the infinite horizon $[0, \infty)$ for any admissible perturbations.*

6 Stability of the Feedback Control System

Let us consider the closed-loop control system

$$\begin{cases} \dfrac{\partial y}{\partial t} + Ay = w(t) \quad \text{a.e. in} \quad Q \\ y(0, x) = 0, \quad x \in \Omega \\ (\alpha y + \dfrac{\partial y}{\partial \nu_A})|_{\Sigma} = u(y(t, x_0)) \end{cases} \tag{6.1}$$

where $u = u(y)$ is the three-positional feedback controller defined in (5.1). Note that although the parabolic equation in (6.1) is linear, the closed-loop system (6.1) is highly *nonlinear* with respect to the state y due to discontinuity of the feedback control law (5.1).

One of the most important characteristics of closed-loop dynamical systems is their *stability* in the sense of maintaining the initial stationary regime after termination of all the perturbations. Such a stability is an obligatory condition for a normal functioning of any automatic control system. We are going to consider the nonlinear control system (6.1) from this viewpoint.

Note that (6.1) is a distributed parameter system where controls acting in the mixed boundary conditions are formed by the current intermediate state $y(t, x_0)$. This generates an *inertia* of the control system and essentially affects its stability. One can easily see that if $y(t, x_0)$ is strictly inside of the dead region $[-\underline{\sigma}, \bar{\sigma}]$ at the time t_0 of termination of all the perturbations, then system (6.1) maintains the stationary regime $y_0(x) \equiv 0$ as $t \to \infty$. This means the stability *in the small* of the initial state $y = 0$ that is *not sufficient* for a normal functioning of the nonlinear control system (6.1) since it does not exclude *self-vibrating regimes.*

Complications may arise when $y(t, x_0)$ reaches the boundary of the dead region if the latter is not sufficiently wide. Indeed, in such cases the transient trajectory moves back and forth between the dead region boundaries under switching control positions in (5.1) with no external perturbations $w(\cdot)$. The next theorem provides effective conditions that exclude such an auto-oscillation and thus ensures the required stability of the closed-loop control system (6.1). The proof of this theorem is based on a *variational approach* to stability that is possible due to monotonicity properties of the parabolic dynamics; see [11, 14] for more details.

Theorem 6 *The closed-loop system (6.1), (5.1) with arbitrary control parameters* $(\bar{u}, \underline{u}, \bar{\sigma}, \underline{\sigma})$ *is stable if*

$$\bar{\sigma} + \underline{\sigma} \geq \min\{\bar{u}, \underline{u}\}\left(\frac{\nu_1 \varphi_1(x_0)}{\lambda_1} - \rho\right) > 0. \tag{6.2}$$

Furthermore, let $\bar{b} \leq \underline{b}$ *and let* $(\bar{\sigma}_1, \underline{\sigma}_1)$ *be computed by*

$$\bar{\sigma}_1 = \bar{b}\gamma - \frac{\nu_1 \varphi_1(x_0)(\bar{b}\gamma - \eta)}{\rho} \quad \textit{and} \quad \underline{\sigma}_1 = \underline{b}\gamma - \frac{\nu_1 \varphi_1(x_0)(\underline{b}\gamma - \eta)}{\rho},$$

i.e., they are the first-term approximations of the asymptotically optimal dead region bounds in (5.2) *and* (5.3). *Then the stability condition* (6.2) *can be written as*

$$2\bar{\sigma}_1 + \underline{\sigma}_1 \geq \eta.$$

References

[1] Agmon, S. (1965) *Lectures on Elliptic Boundary Value Problems.* Van Nostrand, Princeton.
[2] Basar, T. and Bernard, P. (1991) *H_∞-Optimal Control and Related Minimax Design Problems.* Birkhäuser, Boston.
[3] Ioffe, A.D. and Tikhomirov, V.M. (1979) *Theory of Extremal Problems.* North-Holland, Amsterdam.
[4] Friedman, A. (1964) *Partial Differential Equations of Parabolic Type.* Prentice-Hall, Englewood Cliffs, N.J.
[5] van Keulen, B. (1993) *H_∞-Control for Distributed Parameter Systems: A State-Space Approach.* Birkhäuser, Boston.

[6] Krasovskii, N.N. and Subbotin, A.I. (1988) *Game-theoretical Control Problems*. Springer-Verlag, New York.

[7] Lions, J.L. (1971) *Optimal Control of Systems Governed by Partial Differential Equations*. Springer-Verlag, Berlin.

[8] Mackenroth, U. (1982) Convex parabolic boundary control problems with pointwise state constraints. *J. Math. Anal. Appl.*, **87**, 256–277.

[9] Mordukhovich, B.S. (1986) Optimal control of ground water regime on two-way engineering reclamation systems. *Water Resources* **12**, 244–253.

[10] Mordukhovich, B.S. (1988) *Approximation Methods in Problems of Optimization and Control*. Nauka, Moscow.

[11] Mordukhovich, B.S. (1989) Minimax design for a class of distributed control systems. *Autom. Remote Control* **50**, 1333–1340.

[12] Mordukhovich, B.S. and Zhang, K. (1995) Feedback boundary control of constrained parabolic equations in uncertainty conditions. *Proc. 3rd Europ. Cont. Conf.*, Rome, Italy, 129–134.

[13] Mordukhovich, B.S. and Zhang, K. (1997) Minimax control of parabolic systems with Dirichlet boundary conditions and state constraints. *Appl. Math. Optim.*, **36**, 323–360.

[14] Mordukhovich, B.S. and Zhang, K. (1996) Feedback suboptimal control for constrained parabolic systems. Preprint, Wayne State University.

[15] Pontryagin, L.S., Boltyanskii, B.G., Gamkrelidze, R.V. and Mishenko, E.F. (1962) *The Mathematical Theory of Optimal Processes*. Wiley-Interscience, New York.

DYNAMICAL SYSTEMS

9 TWO APPROACHES TO VIABILITY AND SET-MEMBERSHIP STATE ESTIMATION

A.B. KURZHANSKI

Faculty of Computational Mathematics and Cybernetics (VMK), Moscow State University, Moscow 119899, Russia

In this paper we deal with two schemes for the calculation of information (consistency) sets in the guaranteed state estimation problem [5], [6], [14], [12], as well as of viability kernels, [1], [13], with the aim of presenting solution schemes that appear to be somewhat different from those introduced earlier. Here we deal with systems restricted by magnitude constraints. The resulting schemes produce ellipsoidal approximations for the sets required.

Let us start with a linear n-dimensional system

$$\dot{x} = A(t)x + u(t), \quad x(t_0) = x^0 \tag{1}$$

under constraints

$$u(t) \in \mathcal{E}(p(t), P(t)), \; x^0 \in \mathcal{E}(\bar{x}^0, X^0) \tag{2}$$

on $u(t), x^0$ and $t \in [t_0, t_1]$.

In what follows, $(\cdot, \cdot)$ denotes the scalar product in R^n. Symbol

$$\mathcal{E}(a, Q) = \{x : (x - a, Q(x - a)) \leq 1\},$$

where Q is a positive definite matrix ($Q > 0$), stands for an ellipsoid. Admissible $u(t)$ are measurable and the constraint on $u(t)$ is to be taken for almost all t. Functions $p(t)$ and $P(t)$ are measurable and bounded.

We shall begin with a discussion of *dynamic programming techniques for the viability problem.* Consider system (1), (2) under an additional *viability constraint*:

$$x(t) \in \mathcal{E}(q(t), Q(t)), \quad t_0 \leq t \leq t_1, \tag{3}$$

where $q(t)$ and $Q(t)$ are continuous. This constraint may arrive due to a *measurement equation*

$$y(t) = G(t)x + v(t), \quad v(t) \in \mathcal{E}(q(t), Q(t)),$$

where $y(t)$ is the observed measurement output.

We shall look for the *viability kernel* $W[\tau]$ at given instant τ which is the set of all points $x = x(\tau)$ for each of which there exists a control $u = u(t)$ that ensures the viability constraint:

$$x[t] = x(t, t_0, x) \in \mathcal{E}(q(t), Q(t)), \quad \tau \leq t \leq t_1.$$

We shall now determine $W[\tau]$ as *the level set*

$$W[\tau] = \{x : V_v(\tau, x) \leq 1\}$$

for *the viability function* $V_v(\tau, x)$, which we define as the solution to the following problem:

$$V_v(\tau, x) = \min_{u(\cdot)}\{\Phi(\tau, u(\cdot)) | x[t] = x(t, \tau, x | u(\cdot))\}$$

where

$$\Phi(\tau, u(\cdot)) = \max\{J_0, J_1, J_2\},$$

and

$$J_0(x[t_1]) = (x[t_1] - q(t_1), Q(t_1)(x[t_1] - q(t_1))),$$
$$J_1(\tau, u(\cdot)) = \text{esssup}(u(t) - p(t), P(t)(u(t) - p(t)))),,$$
$$J_2(\tau, x[t]) = \max(x[t], Q(t)x[t])$$

with $t \in [\tau, t_1]$ and $x[t] = x(t, \tau, x | u(\cdot))$ being the trajectory of system (1) that starts at position $\{\tau, x\}$ and is steered by control $u(t)$.

The solution to this problem may be described by a certain "forward" dynamic programming (H-J-B) equation, [2]. In order to avoid generalized solutions of this equation, we shall follow the scheme of the latter paper by solving a linear-quadratic extremal problem which is to minimize

$$\Lambda(\tau, x, u(\cdot), \omega(\cdot)) = \int_{\tau}^{t_1} (\gamma(t)(x[t], Q(t)x[t]) + \beta(t)(u(t) - p(t), P(t)(u(t) - p(t)))dt +$$
$$+ \alpha(x[t_1], Q(t_1)x[t_1])$$

over $u(\cdot)$, with $x[t] = x(t, \tau, x | u(\cdot))$. Here $\omega(\cdot) = \{\alpha, \beta(\cdot), \gamma(\cdot))\}$,

$$\alpha > 0, \ \beta(t) > 0, \ \gamma(t) > 0, \quad \alpha + \int_{\tau}^{t_1} (\beta(t) + \gamma(t))dt = 1,$$

$\beta(\cdot)$ is measurable and bounded, and $\gamma(\cdot)$ is continuous. The variety of such elements $\omega(\cdot)$ is further denoted as Ω.

Then, in analogy with [2], we have

$$V_v(\tau, x) = \min_{u(\cdot)}\{\Phi(\tau, u(\cdot)) | x[t] = x(t, \tau, x | u(\cdot))\}$$
$$= \min_{u(\cdot)} \sup_{\omega(\cdot)} \Lambda(\tau, x, u(\cdot), \omega(\cdot)).$$

Here the structure of function Λ is such that operations of min and sup may be interchanged ([3]). Doing this, we denote

$$V_v(\tau, x) = \sup_{\omega(\cdot)} V(\tau, x, \omega),$$

where

$$V(\tau, x, \omega(\cdot)) = \min_{u(\cdot)} \Lambda(\tau, x, u(\cdot), \omega(\cdot)).$$

We again look for this function as a quadratic form

$$V(\tau, x, \omega) = (x - z(\tau, \gamma(\cdot)), \mathcal{P}(\tau, \omega(\cdot))(x - z(\tau, \gamma(\cdot))), \tag{4}$$

where $\mathcal{P}[t] = \mathcal{P}(t, \omega(\cdot)), z[t] = z(t, \gamma(\cdot)), k = k(t, \gamma(\cdot))$ satisfy equations

$$\dot{\mathcal{P}} = -\mathcal{P}A(t) - A'(t)\mathcal{P} + \beta^{-1}(t)\mathcal{P}P(t)\mathcal{P} - \gamma(\cdot)Q^{-1}(t), \tag{5}$$

$$\dot{z} = A(t)z - \gamma(t)\mathcal{P}^{-1}Q^{-1}(t)(z + q(t)) + p(t), \tag{6}$$

$$k^2(t) = -\gamma(\cdot)(z + q(t), Q^{-1}(t)(z + q(t))), \tag{7}$$

$$\mathcal{P}(t_1) = \alpha Q^{-1}(t_1),\ z(t_1) = q(t_1),\ k(t_1) = 0. \tag{8}$$

It may be more convenient to deal with matrix $X_v(t) = \mathcal{P}^{-1}[t]$ however, which satisfies equation

$$\dot{X}_v = A(t)X_v + X_vA'(t) + \gamma(\cdot))X_vQ^{-1}(t)X_v - \beta^{-1}(t)P(t), \tag{9}$$

$$X_v(t_1) = \alpha^{-1}Q(t_1). \tag{10}$$

Summarizing the given manipulations we formulate the following assertion.

Lemma 1 *The viability function $V_v(\tau, x)$ is the upper envelope*

$$V_v(\tau, x) = \sup\{V(\tau, x, \omega(\cdot))|\omega(\cdot) \in \Omega\}$$

of a parametrized variety of quadratic forms $V(\tau, x, \omega(\cdot))$ of type (4) over the functional parameter $\omega(\cdot) = \{\alpha, \beta(\cdot), \gamma(\cdot))\}$, where $\omega(\cdot) \in \Omega$.

Since the level sets for $V_v(\tau, x, \omega(\cdot))$ are ellipsoids, namely

$$W[\tau, \omega(\cdot)] = \mathcal{E}(z[\tau], (1 - k^2[\tau])^{-1}X[\tau])$$

and since $W[\tau]$ is a level set for $V_v(\tau, x)$, we are able, due to the previous lemma, to come to

Theorem 1 *The viability set $W[\tau]$ is the intersection of ellipsoids, namely*

$$W[\tau] = \{\cap\mathcal{E}(z[\tau], (1 - k^2[\tau])^{-1}X[\tau])|\omega(\cdot) \in \Omega\},$$

where $\mathcal{P}, z, k, X_v$, are defined through equations (5)–(10).

The set-valued function $W[t]$, $\tau \leq t \leq t_1$ is known as *the viability tube* which may thus be approximated by ellipsoidal-valued tubes along the schemes given here.

A similar dynamic programming scheme allows to be applied to the calculation of information (consistency) sets for the set-membership (bounding) approach to the state-estimation problem. It is important to emphasize that the dynamic programming schemes show close connections with the approaches to uncertain systems based on Liapunov functions [11].

As is rather obvious, these information domains are nothing else than *attainability domains under state constraints* when the last are given, for example, by relations of type (3). These domains $\mathcal{X}(\tau)$ may be described through Dynamic Programming similarly to the above. However, some other types of ellipsoidal estimates and their dynamics may be derived for $\mathcal{X}(\tau)$ directly, through the funnel equations of papers [10], [8] and some "elementary" formulae of paper [9].

We shall consider the attainability domain $\mathcal{X}[\tau]$ for system

$$\dot{x} = u(t), \tag{11}$$

under constraints (2) on $u(t), x^0$ and state constraint

$$x(t) \in \mathcal{E}(y(t), K(t)), \tag{12}$$

where the matrix-valued function $K(t) > 0$ $(K(t) \in \mathcal{L}(R^n, R^n))$ and the function $y(t) \in R^n$ (the observed output in the state estimation problem) are assumed to be continuous[1].

Note that in system (1) we may set $A(t) \equiv 0$ without loss of generality, provided $p(t), P(t)$ are time-dependent.

We now follow the techniques of funnel equations with set-valued solutions, described in papers [8], [10]. For the attainability domain $\mathcal{X}(t)$ under state constraint (12) this gives

$$\mathcal{X}(t+\sigma) = [\mathcal{X}(t) + \sigma\mathcal{E}(p(t), P(t) + f(t)) \cap \mathcal{E}(y(t+\sigma), K(t+\sigma)] + o(\sigma), \quad \sigma > 0.$$

Presuming $\mathcal{X}(t) = \mathcal{E}(x(t), X(t))$, we shall seek for the external ellipsoidal estimate $\mathcal{E}(x(t+\sigma), X(t+\sigma))$ of $\mathcal{X}(t+\sigma)$. To do this we shall use relations similar to those of papers [8], [4].

Namely, using these relations, we first take the estimate

$$\mathcal{E}(x(t), X(t)) + \sigma\mathcal{E}(p(t), P(t) + f(t)) \subset \mathcal{E}(\tilde{x}(t), \tilde{X}(t)),$$

where $\tilde{x}(t) = x(t) + \sigma p(t)$, and

$$\tilde{X}(t) = (1+q)X(t) + (1+q^{-1})\sigma^2 P(t), \quad q > 0. \tag{13}$$

---

[1] The case of measurable functions $y(t)$ which allows more complicated discontinuities in $y(t)$ and is of special interest in applications is treated in paper [4].

Further on, we have

$$\mathcal{E}(\tilde{x}, \tilde{X}) \cap \mathcal{E}(y, K) \subset \mathcal{E}(x(t+\sigma, X(t+\sigma)),$$

where

$$x(t+\sigma) = (I - M)(x(t) + \sigma p(t)) + My(t+\sigma), \tag{14}$$

and

$$X(t+\sigma) = (1+\pi)(I-M)\tilde{X}(t)(I-M)' + (1+\pi^{-1})MK(t+\sigma)M', \ \pi > 0. \tag{15}$$

Making the substitutions

$$q = \sigma\bar{q}, \ \pi = \sigma\bar{\pi}, = \sigma\bar{M},$$

collecting (13)–(15) together and leaving the terms of order ≤ 1 in σ, we come to

$$x(t+\sigma) - x(t) = \sigma p + \sigma\bar{M}(y(t+\sigma) - x(t))$$

and also

$$X(t+\sigma) - X(t) = \sigma[(\bar{\pi} + \bar{q})X(t) - \bar{M}X - X\bar{M}' + \bar{q}^{-1}P + \bar{\pi}^{-1}\bar{M}K(t+\sigma)\bar{M}'].$$

Dividing both parts of the previous equations by $\sigma > 0$ and passing to the limit $\sigma \to +0$, we further come, in view of the continuity of $y(t), K(t)$, to differential equations (deleting the bars in the notations)

$$\dot{x} = p(t) + M(t)(y(t) - x(t)), \tag{16}$$

$$\dot{X} = (\pi(t) + q(t))X + q(t)^{-1}P - M(t)X - XM'(t) + \pi^{-1}M(t)K(t)M'(t), \tag{17}$$

where

$$x(t_0) = x_0, \quad X(t_0) = X_0, \tag{18}$$

and $\pi(t) > 0, q(t) > 0, M(t)$ are continuous functions.

What further follows is the assertion

Theorem 2 *The attainability domain $\mathcal{X}(\tau)$ for system (11) under restrictions (2) and state constraints (12) (with $y(t), K(t)$ continuous) satisfies the inclusion $\mathcal{X}(\tau) \in \mathcal{E}(x(\tau), X(\tau))$, where $x(t), X(t)$ satisfy the differential equations (16), (17) within the interval $t_0 \leq t \leq \tau$, and the boundary conditions (18).*

Moreover, the following relation is true:

$$\mathcal{X}(\tau) = \cap\{\mathcal{E}(x(t), X(t)) | \pi(\cdot), q(\cdot), M(\cdot)\} \tag{19}$$

where $\pi(t) > 0, q(t) > 0, M(t)$ are continuous functions.

Suppose $A(t) \not\equiv 0$ and the state constraint (22) is substituted by relation

$$G(t)x(t) \in \mathcal{E}(y(t), K(t))$$

where $y(t) \in R^m$, $K(t) \in \mathcal{L}(R^m, R^m)$, $G(t) \in \mathcal{L}(R^n, R^m)$, and $G(t)$ is continuous. Then the previous relations together with Theorem 2 are still true with obvious changes. Namely, (16), (17) should be substituted by

$$\dot{x} = A(t)x + p(t) + M(t)(y(t) - G(t)x), \tag{20}$$

$$\begin{aligned} \dot{X} = (A(t) - M(t)G(t))X + X(A(t) - G'(t)M'(t)) + \\ + (\pi(t) + q(t))X + q(t)^{-1}P(t) + \pi^{-1}M(t)K(t)M'(t), \end{aligned} \tag{21}$$

with same boundary conditions (18).

It is not difficult to observe that the variety of ellipsoids given in (19) depends on more parameters than in the dynamic programming technique and is therefore a "richer" variety. It is then natural to expect that in the process of selecting an optimal ellipsoid (relative to an appropriate preassigned criterion) it will produce a "tighter" optimal ellipsoid than the dynamic programming technique.

System (20), (21) was derived under the assumption that function $y(t)$ is continuous. However, we may as well assume that $y(t)$ is allowed to be piece-wise "right"-continuous. Then the respective value $y(t)$ should be taken as $y(t) = y(t+0)$.

References

[1] Aubin, J.-P. (1991) *Viability Theory*. Birkhäuser, Boston.

[2] Baras, J.S. and Kurzhanski, A.B. (1995) Nonlinear filtering: the set-membership (bounding) and the H_∞ approaches. *Proc. of the IFAC NOLCOS Conference*, Tahoe, CA USA, Plenum Press.

[3] Fan, K.Y. (1953) Minmax theorems. *Proc. Nat. Acad. of Sci. USA*, **39**(1), 42–47.

[4] Filippova, T.F., Kurzhanski, A.B., Sugimoto, K. and Vályi, I. (1993) Ellipsoidal calculus, singular perturbations and the state estimation problem for uncertain systems, in *IIASA Working Paper, Laxenburg*, WP-92-51, (see also *JMSEC, N2, 1996*).

[5] Kurzhanski, A.B. (1972) Differential games of observation. *Sov. Math. Doklady*, **13**(6), 1556–1560.

[6] Kurzhanski, A.B. (1977) *Control and Observation Under Uncertainty*, Nauka, Moscow.

[7] Kurzhanski, A.B. and Filippova, T.F. (1993) On the theory of trajectory tubes — a mathematical formalism for uncertain dynamics, viability and control, in *Advances in Nonlinear Dynamics and Control: a Report from Russia*, A.B. Kurzhanski, ed., ser. PSCT 17, Birkhäuser, Boston, 122–188.

[8] Kurzhanski, A.B., Sugimoto, K. and Valyi, I. (1994) Guaranteed state estimation for dynamic systems: ellipsoidal techniques, *Intern. Journ. of Adaptive Contr. and Sign. Proc.*, **8**, 85–101.

[9] Kurzhanski, A.B. and Vályi, I. (1988) Set valued solutions to control problems and their approximation, in: A. Bensoussan, J. L. Lions Eds., *Analysis and Optimization of Systems*, Lecture Notes in *Control and Information Systems*, Vol. 111, 775–785. Springer Verlag.

[10] Kurzhanski, A.B. and Vályi, I. (1991) Ellipsoidal techniques for dynamic systems: the problems of control synthesis. *Dynamics and Control*, **1**, 357–378.

[11] Leitmann, G. (1993) One approach to the control of uncertain dynamical systems. *Proc. 6-th Workshop on Dynamics and Control*, Vienna.
[12] Milanese, M. and VICINO A. (1991) Optimal estimation for dynamic systems with set-membership uncertainty: an overview. *Automatica*, **27**, 997–1009.
[13] Saint-Pierre, P. (1994) Approximation of the viability kernel. *Applied Math. and Optim.*, **29**, 187–209.
[14] Scheweppe, F.C. (1973) *Uncertain Dynamic Systems*. Prentice Hall, Englewood Cliffs, NJ.

10 TIME CONSTANTS FOR THE QUADRATIC LYAPUNOV-FUNCTIONS

LÁSZLÓ ÁKOS

Dept. of Operations Research, Eötvös University, H-1088 Budapest, Hungary

1 Introduction

Let us consider a linear time invariant dynamic system described by the state equation

$$\frac{dx(t)}{dt} = Ax(t) + Bu(t), \tag{1}$$

where x is the state vector, u is the input vector, and (A, B) is controllable, $rank(B) = m$. Similar equations may occur, for example, via linearization in robotics [2], where the meaning of x is an error or deviation from ideal, and error estimates are demanded for the controlled systems. Let the matrix K be any stabilizing constant matrix, i.e. $A + BK$ is stable. We assume, that a Lyapunov-function is defined as

$$L(x(t)) = x^*(t)Px(t), \tag{2}$$

where P is the solution of the equation

$$P(A + BK) + (A + BK)^*P + Q = 0, \tag{3}$$

with a given positive definite Q. The latter is an important property in application of Lyapunov's method, because in the equations describing real physical processes uncertainties may occur in the modeled parameters, or unmodeled dynamics. If the matrix Q were semi-definite, even small disturbances could make it indefinite. Suppose that we have a positive number α, for which

$$Q \geq \alpha P. \tag{4}$$

The time derivative of $L(x(t))$ along the trajectory of system (1) with the state feedback $u(t) = Kx(t)$ then satisfies the inequality

$$\frac{dL(t)}{dt} = -x^*Qx \leq -\alpha L(x),$$

and it can be seen easily that

$$L(x(t)) \leq L(x(0))e^{-\alpha t}.$$

The number $\frac{1}{\alpha}$ can be stated as a time constant; it gives the speed of convergence. Such methods are well known [3], but not effective. Our goal is to prove that for fixed matrices A, B, Q there is a finite supremum $\hat{\alpha}$ of α — achievable by these methods — in the system (1)–(4). These computations lead to the following problem: find the supremum of the numbers α for all free P and K satisfying (3), (4) such that $A + BK$ is stable. It will be shown that the upper bound

$$\hat{\alpha} \leq cond_2(T^* Q T)$$

is valid if $n > m$, where T is any one of the matrices transforming (A, B) to the "Luenberger" canonical form. (See Theorem 1 below.)

The fact that we have got an upper bound is at first surprising: if the pair (A, B) is controllable, then any $\lambda_1, \ldots, \lambda_n$ complex eigenvalues having negative real parts with large absolute value may be realized for the closed loop system $\frac{dx}{dt} = (A + BK)x$. We still wouldn't find a matrix K which guarantees with the method described above a better convergence in $L(x(t))$ than $L(x(0))e^{-\hat{\alpha} t}$.

We will consider the following generalization.

Let Q_1 be a given positive semi-definite matrix, Q_2 positive definite matrix, such that the pair (Q_1, A) is detectable, and let K and P satisfy

$$P(A + BK) + (A + BK)^* P + Q_1 = 0, \tag{5}$$

$$Q_2 \geq \alpha P. \tag{6}$$

$$\hat{\alpha} = \sup_{P>0,\ K \in S_{(A,B)}} \alpha = ?\ ,$$

where $S_{(A,B)} = \{K : \ A + BK \text{ is asymptotically stable}\}$.

We will see that this problem arises naturally when we consider the so-called cheap control problem.

The following result will be shown below: If $\hat{\alpha} = \infty$, then there exists a matrix $C \in R^{m \times n}$, satisfying $C^* C = Q_1$ such that for every positive ϵ there exists a matrix $L \in R^{m \times n}$ and a number $s_0 > 0$ such that $\|L - C\| < \epsilon$, and $A - BLs$ is stable, if $s > s_0$. (See Theorem 2 below.)

2 Preliminaries

Lemma 1 Under the assumptions (5) and (6)

$$\frac{1}{\hat{\alpha}} = \inf_{K \in S_{(A,B)}} \left\| \int_0^\infty Q_2^{-\frac{1}{2}} e^{(A+BK)^* t} Q_1 e^{(A+BK)t} Q_2^{-\frac{1}{2}} dt \right\|$$

$$= \sup_{x(0)=Q_2^{-\frac{1}{2}}v, \|v\|=1} \left(\inf_{K \in S_{(A,B)}} \int_0^\infty x^*(t) Q_1 x(t) dt \right);$$

where $x(t)$ is the solution of

$$\frac{dx(t)}{dt} = Ax(t) + Bu(t).$$

Proof Equation (5) can be solved since $(A + BK)$ is stable. Using (6) it can be seen easily that

$$Q_2^{-\frac{1}{2}} \int_0^\infty e^{(A+BK)^* t} Q_1 e^{(A+BK)t} dt Q_2^{-\frac{1}{2}} \leq \frac{1}{\alpha} I,$$

and this implies the first equation.

The second equation follows trivially.

We assume that a parametrized quadratic cost function related to equation (1) is defined as:

$$J_\eta(x_0) = \inf_{u \in L^2} \int_0^\infty (x^*(t) Q_1 x(t) + \eta u^*(t) u(t)) dt \tag{7}$$

(with $\frac{dx}{dt}(t) = Ax(t) + Bu(t)$).

The asymptotic optimal solution in the case when the number η is tending to zero is called the cheap control problem (see, for example, [4] or [5]). The next lemmas give the connection between our problem and the cheap control problem.

Lemma 2 If the pair (Q_1, A) is detectable, and (A, B) is stabilizable, then

$$J_0(x_0) = \lim_{\eta \to 0} J_\eta(x_0) = \lim_{\eta \to 0} x_0^* P_\eta x_0$$

holds; where P_η is the maximal symmetric solution of the algebraic Riccati equation

$$A^* P + PA + Q_1 - \frac{1}{\eta} PBB^* P = 0. \tag{8}$$

Proof Since we may write $J_0(x_0) \leq J_\eta(x_0) \leq J_\mu(x_0)$ for every $0 \leq \eta \leq \mu$ (see [1]), it is convenient to assume that $J_0(x_0) < J_\eta(x_0)$. In this case there exist an $\epsilon > 0$ and a square integrable control $\hat{u}$ such that for the solution $x_{\hat{u}}$ of (1)

$$\int_0^\infty x_{\hat{u}}^*(t) Q_1 x_{\hat{u}}(t) dt + \epsilon < J_\eta(x_0)$$

is valid for all $\eta > 0$. $J_\eta(x_0)$ is the value of the minimum, and therefore

$$J_\eta(x_0) \leq \int_0^\infty (x_{\hat{u}}^*(t) Q_1 x_{\hat{u}}(t) + \eta \hat{u}^* \hat{u}) dt.$$

From the two inequalities it follows immediately that

$$\epsilon \leq \eta \int_0^\infty \hat{u}^* \hat{u} dt$$

for every $\eta > 0$, and this is a contradiction.

Remark Consider the matrix

$$P_0 = \lim_{\eta \to 0} P_\eta.$$

The lemmas imply immediately the equation

$$\frac{1}{\hat{\alpha}} = \lambda_{max}\left(Q_2^{-\frac{1}{2}} P_0 Q_2^{-\frac{1}{2}}\right) = \rho(P_0 Q_2^{-1}).$$

Lemma 3 Define

$$A = \begin{bmatrix} 0 & 1 \\ 0 & 0 \end{bmatrix} \quad B = \begin{bmatrix} 0 \\ 1 \end{bmatrix} \quad Q_1 = q \begin{bmatrix} 1 & 0 \\ 0 & 1 \end{bmatrix}.$$

where $q > 0$. In this case

$$P_0 = q \begin{bmatrix} 1 & 0 \\ 0 & 0 \end{bmatrix}.$$

Proof The maximal solution of equation 8 is

$$P_\eta = \begin{bmatrix} \sqrt{q^2 + 2q\sqrt{rq}} & \sqrt{rq} \\ \sqrt{rq} & \sqrt{qr + 2r\sqrt{rq}} \end{bmatrix},$$

and

$$P_0 = \lim_{\eta \to 0} P_\eta.$$

Lemma 4 If we replace the matrix A by $A - BL$ in (3) or (5), where L is any $m \times n$ matrix, then the value $\hat{\alpha}$ does not change.

Proof From the proof of Lemma 1 we can see that there exists a series of feedbacks $\{K_i\}_{i=0}^\infty$ such that with the solution of

$$\frac{dx_{K_i}(t)}{dt} = (A + BK_i) x_{K_i}(t)$$

the relation

$$\frac{1}{\hat{\alpha}} = \sup_{x(0)=Q_2^{-\frac{1}{2}}v, \|v\|=1} (\lim_{i\to\infty}) \int_0^\infty x_{K_i}^*(t) Q_1 x_{K_i}(t) dt$$

holds. (For instance $K_i = -iB^* P_{\frac{1}{i}}$.) In the new problem the series can be replaced by $\{K_i - L\}_{i=0}^\infty$, and the costs remain unchanged.

3 Main Results

Now we are able to prove Theorem 1. Assume that $Q_1 = Q_2 = Q > 0$, the pair (A, B) is controllable, and the size of the matrices are $n \times n$ and $n \times m$ respectively, $rank(B) = m < n$.

Theorem 1 *Consider equation (3) and inequality (4). The upper bound*

$$\hat{\alpha} \le cond_2(T^* Q T)$$

holds, where T is any one of the matrices transforming (A, B) to the "Luenberger" canonical form.

Proof Define

$$\bar{A} = T^{-1}AT, \quad \bar{B} = T^{-1}B.$$

The construction of the transformation T is very complicated. In our proof it suffices to note that $\bar{B}$ contains m columns of an $n \times n$ identity matrix, and $\bar{A}$ has nonzero entries only in the rows in which $\bar{B}$ has nonzero entries, or — in other rows — only above the main diagonal, where they are just equal to one. For instance

$$\bar{A} = \begin{bmatrix} 0 & 1 & 0 & 0 & 0 \\ 0 & 0 & 1 & 0 & 0 \\ a_3 & a_2 & a_1 & b_2 & b_1 \\ 0 & 0 & 0 & 0 & 1 \\ c_2 & c_1 & 0 & d_2 & d_1 \end{bmatrix} \quad \bar{B} = \begin{bmatrix} 0 & 0 \\ 0 & 0 \\ 1 & 0 \\ 0 & 0 \\ 0 & 1 \end{bmatrix},$$

where $a_1, a_2, a_3, b_1, b_2, c_1, c_2, d_1, d_2$ are real numbers. Let $\bar{L}$ be the $m \times n$ matrix which contains all the rows of the matrix $\bar{A}$ corresponding the nonzero rows of $\bar{B}$; let $L = \bar{L}T^{-1}$. All eigenvalues of the matrices $\bar{A}_I := \bar{A} - \bar{B}\bar{L}$, and $A_I := T(\bar{A} - \bar{B}\bar{L})T^{-1} = A - B\bar{L}T^{-1} = A - BL$ are zero, since $\bar{A}_I$ is a Jordan normal form with zeros in the diagonal. By Lemma 4 the matrix A_I gives the same value $\hat{\alpha}$ as the matrix A. If we replace the matrix A by A_I, and define $y(\cdot) = T^{-1}x(\cdot)$, Lemma 1 implies that

$$\frac{1}{\hat{\alpha}} = \sup_{y(0)=T^{-1}Q^{-\frac{1}{2}}v, \|v\|=1} \left(\inf_{K\in S_{(A,B)}} \int_0^\infty y^*(t)\bar{Q}y(t)dt \right);$$

where $\frac{dy(t)}{dt} = (\bar{A}_I + \bar{B}K)y(t)$, and $\bar{Q} = T^*QT$. Note that $\lambda_{min}(\bar{Q}) > 0$, since Q is positive definite and T is nonsingular. The matrix has nonzero entries above the diagonal, since $m < n$. Let the i-th row be the first of them which contains a 1, i.e. $(\bar{A}_I)_{i,i+1}$. It is easy to see that

$$\frac{1}{\hat{\alpha}} \geq \lambda_{min}(\bar{Q}) \sup_{y(0)=T^{-1}Q^{-\frac{1}{2}}v, \|v\|=1} \left(\inf_{K \in S_{(A,B)}} \int_0^\infty y^*(t)y(t)dt \right) \geq$$

(where $\frac{dy(t)}{dt} = (\bar{A}_I + \bar{B}K)y(t)$)

$$\geq J := \lambda_{min}(\bar{Q}) \sup_{z(0)=T^{-1}Q^{-\frac{1}{2}}v, \|v\|=1} \left(\inf_{K \in S_{(A,B)}} \int_0^\infty (z_i^2(t) + z_{i+1}^2(t))dt \right)$$

(where $\frac{dz(t)}{dt} = (\bar{A}_I + \bar{B}K)z(t)$). It can be shown from Lemma 3 that

$$\inf_{K \in S_{(A,B)}} \int_0^\infty (z_i^2(t) + z_{i+1}^2(t))dt) \geq z_i^2(0)$$

holds. If $\bar{B}$ has in the i+1-th row a 'one', then the corresponding element of u plays the role of the control in Lemma 3. In the other case the 'control' is $z_{i+1}(\cdot)$; and we have inequality. The inequality reduces to

$$\frac{1}{\hat{\alpha}} \geq \lambda_{min}(\bar{Q}) \sup_{z(0)=T^{-1}Q^{-\frac{1}{2}}v, \|v\|=1} z_i^2(0)$$

We introduce the vector

$$v = \frac{Q^{\frac{1}{2}}Te_i}{\|Q^{\frac{1}{2}}Te_i\|}$$

where e_i is the i-th unit vector. This leads to

$$\frac{1}{\hat{\alpha}} \geq \lambda_{min}(\bar{Q}) \sup_{z(0)=T^{-1}Q^{-\frac{1}{2}}v, \|v\|=1} z_i^2(0) \geq \frac{\lambda_{min}(\bar{Q})}{e_i^*T^*QTe_i} \geq \frac{\lambda_{min}(\bar{Q})}{\lambda_{max}(\bar{Q})};$$

by which the proof is complete:

$$\frac{1}{\hat{\alpha}} \geq \frac{1}{cond_2(T^*QT)}.$$

Theorem 2 *Consider equations (5) and (6). If $\hat{\alpha} = \infty$, then there exists a matrix $C \in R^{m\times n}$, satisfying $C^*C = Q_1$ such that for every positive ϵ there exists a matrix $L \in R^{m\times n}$ and a number $s_0 > 0$ such that $\|L - C\| < \epsilon$, and $A - BLs$ is stable, if $s > s_0$.*

Proof $\hat{\alpha} = \infty$ if and only if

$$P_0 = \lim_{\eta \to 0} P_\eta = 0$$

for the solution of the Algebraic Riccati Equation (8) holds. In this case let $\eta \to 0$ in (8). Then

$$Q_1 = lim_{\eta \to 0} \frac{1}{\eta} P_\eta BB^* P_\eta.$$

We can define the matrix

$$C = lim_{\eta \to 0} \frac{1}{\sqrt{\eta}} B^* P_\eta.$$

If ϵ is an arbitrary positive number, then there exists a number η, such that

$$\left\| C - \frac{1}{\sqrt{\eta}} B^* P_\eta \right\| < \epsilon.$$

Define $L = \frac{1}{\sqrt{\eta}} B^* P_\eta$, $s_0 = \frac{1}{\sqrt{\eta}}$. The matrix $A - BLs = A - \frac{s}{\sqrt{\eta}} BB^* P_\eta$ is stable, because equation (8) gives for $s \geq s_0$

$$\left(A - \frac{s}{\sqrt{\eta}} BB^* P_\eta \right)^* P_\eta + P_\eta \left(A - \frac{s}{\sqrt{\eta}} BB^* P_\eta \right) + Q + \left(\frac{2s}{\sqrt{\eta}} - \frac{1}{\eta} \right) P_\eta BB^* P_\eta = 0,$$

and the stability follows from the standard results for solutions of algebraic Lyapunov equations.

Remark We can see by the theorem that $rang(Q_1) > rang(B)$ implies $\bar{\alpha} < \infty$.

References

[1] Ran, A.C.M. and Vreugdenhil, R. (1989) Existence and comparation theorems for Algebraic Riccati equations for continuous and discrete-time systems. *Linear Algebra and its Applications*, **99**, 63–83.
[2] Zhihua Qu and Dorsey, J. (1991) Robust tracking control of robots by a linear feedback law. *IEEE Transactions on Automatic Control*, **36**, 1081–1084
[3] Kalman, R.E. and Bertram, J.E. (1960) Control system analysis and design via the "second method" of Lyapunov. *Journal of Basic Engineering*, June, 371–393.
[4] Geerts, T. Structure of Linear-Quadratic Control. Dissertation, University of Eindhoven.
[5] Wonham, W. M. (1985) Linear Multivariable Control. A Geometric Approach. Springer-Verlag Applications of Mathematics.

11 QUALITATIVE ANALYSIS WITH RESPECT TO TWO MEASURES FOR POPULATION GROWTH MODELS OF KOLMOGOROV TYPE

ANATOLI A. MARTYNYUK

Department of Stability of Processes, Institute of Mechanics, National Ukrainian Academy of Sciences, 152057 Kiev 57, Ukraine

For unperturbed and perturbed Kolmogorov models of population dynamics, new sufficient conditions of stability and boundedness with respect to two criteria (measures) are provided. An application to a generalized Lotka-Volterra equation is given.

1 Introduction

An approach to treating Kolmogorov models of population dynamics (see [1]) via two measures was suggested in [2–4]. This approach worked out for both ordinary and partial differential equations allows conditions to obtain sufficient for the Kolmogorov models to possess various dynamical properties. The present paper continues the investigation in this direction and provides new stability and boundedness conditions for Kolmogorov-type ordinary differential equations.

In Section 2 unperturbed and perturbed Kolmogorov models are introduced; multiplicative and additive perturbations are taken into account. In Sections 3 and 4 boundedness and stability conditions with respect to two measures are formulated. In Section 5 these conditions are applied for the analysis of a generalized Lotka-Volterra model.

2 Generalized Kolmogorov Model

We consider a system of ordinary differential equations

$$\frac{dx_i}{dt} = \beta_i(x_i)F_i(t, x_1, \ldots, x_n, \mu), \quad x_i(t_0) = x_{i0} \geq 0, \quad i = 1, 2, \ldots, n. \tag{2.1}$$

Here β_i are infinitely differentiable functions on $R_+ = [0, \infty)$, $\beta_i(0) = 0$, $\beta_i'(x_i) > 0$ for $x_i > 0$, $\beta_i^j(x_i) \geq 0$ $(j = 2, 3, \ldots)$, and $F_i \in C(R_+ \times R_+^n \times M^k, R)$, where $M^k = [0, 1] \times \ldots \times [0, 1] (i = 1, 2, \ldots, n)$. System (2.1) represents a multiplicatively and additively perturbed Kolmogorov equation serving for modeling dynamics of populations. Along with (2.1), we consider an unperturbed Kolmogorov system,

$$\frac{dx_i}{dt} = K_i x_i f_i(t, x_1, \ldots, x_n), \quad x_i(t_0) = x_{i0} \geq 0, \quad i = 1, 2, \ldots, n. \tag{2.2}$$

Here $K_i > 0$ and $f_i(t, x_1, \ldots, x_n) = F_i(t, x_1, \ldots, x_n, 0) (i = 1, 2, \ldots, n)$. In what follows, x stands for a vector with coordinates $x_1, \ldots, x_n$.

3 Boundedness with Respect to Two Measures

A qualitative analysis of systems (2.1) and (2.2) will involve two measures of boundedness, ρ and ρ_0. These are elements from

$$\mathcal{M} = \left\{ \rho \in C(R_+ \times R_+^n, R_+) : \inf_{(t,x) \in R_+ \times R_+^n} \rho(t, x) = 0 \right\}.$$

Definition 3.1 The solutions of system (2.2) are said to be

(i) (ρ_0, ρ)-bounded if there exists a nonnegative function $\delta(t_0, \alpha)$ on R_+^2 continuous in t_0, satisfying $\delta(t_0, \alpha) > 0$ for $\alpha > 0$, and such that for every $\alpha > 0$ and every $(t_0, x_0) \in R_+ \times R_+^n$, solution $x(t)$ of (2.2) satisfies

$$\rho(t, x(t)) < \delta(t_0, \alpha) \quad \text{for all} \quad t \geq t_0$$

whenever $\rho_0(t_0, x_0) < \alpha$;

(ii) uniformly (ρ_0, ρ)-bounded if $\delta(t_0, \alpha)$ in (i) does not depend on t_0.

Definition 3.2 The solutions of system (2.1) are said to be

(i) (ρ_0, ρ, μ)-bounded if there exist nonnegative functions $\delta(t_0, \alpha)$ and $\mu^*(t_0, \alpha)$ on R_+^2 such that $\delta(t_0, \alpha)$ is continuous in t_0, $\delta(t_0, \alpha) > 0$, $\mu^*(t_0, \alpha) > 0$ for $\alpha > 0$, and for every $\epsilon > 0$ and every $(t_0, x_0) \in R_+ \times R_+^n$, solution $x(t, \mu)$ of (2.1) satisfies

$$\rho(t, x(t, \mu)) < \delta(t_0, \alpha) \quad \text{for all} \quad t \geq t_0$$

whenever $\rho_0(t_0, x_0) < \alpha$ and $\mu_i < \mu^*(t_0, \alpha)$, $i = 1, 2, \ldots, k$.

(ii) uniformly (ρ_0, ρ)-bounded if $\delta(t_0, \alpha)$ in (i) does not depend on t_0.

3.1 Unperturbed Kolmogorov System

We shall provide a (ρ_0, ρ)-boundedness criterion for system (2.2) using the Lyapunov's direct method. Assume the following definitions.

Definition 3.3 A continuous function $a : R_+ \mapsto R_+$ is said to belong to

(i) class $\mathcal{KR}$ if a is strictly increasing and $\lim_{r\to\infty} a(r) = \infty$,
(ii) class $\mathcal{K}$ if a is strictly increasing and $a(0) = 0$.

Definition 3.4 Let $\rho, \rho_0 \in \mathcal{M}$.We say that

(i) ρ is continuous with respect to ρ_0 if there exist $\Delta > 0$ and $\phi \in C(R^2_+, R)$ such that $s \mapsto \phi(t, s)$ belongs to class $\mathcal{KR}$ for each $t \in R_+$, and $\rho(t, x) < \phi(t, \rho_0(t, x))$ holds whenever $\rho_0(t, x) < \Delta$,
(ii) ρ is uniformly continuous with respect to ρ_0 if $\phi(t, s)$ in (i) does not depend on t.

Consider a matrix function

$$U(t, x) = [u_{ij}(t, x)], \quad i, j = 1, 2, \ldots, m$$

where $u_{ij} \in C(R_+ \times R^n_+, R)$. Take $y \in R^m$ and introduce a Lyapunov function

$$V(t, x) = y^T U(t, x) y \tag{3.1}$$

(see [5]). Define the derivative of V along the vector field (2.2) at point (t, x) by

$$D^+V(t, x)\mid_{(2.2)} = \limsup_{\theta\to 0+}[V(t + \theta, x + \theta \operatorname{diag}(Kx)f(t, x)) - V(t, x)]\theta^{-1}.$$

Here $\operatorname{diag}(Kx)$ is a diagonal $n \times n$ matrix with $K_1x_1, \ldots, K_nx_n$ on the diagonal. The next theorem formulates a sufficient condition for (ρ_0, ρ)-boundedness in terms of the Lyapunov function V.

Theorem 3.1 *Suppose that*

(1) $\rho, \rho_0 \in \mathcal{M}$ and ρ is continuous with respect to ρ_0,
(2) $V(t, x)$ is locally Lipschitz in x, $a \in \mathcal{KR}$, $w \in C(R^2_+, R_+)$, and

$$a(\rho(t, x)) \le V(t, x) \le w(t, \rho_0(t, x))$$

for all $(t, x) \in R_+ \times R^n_+$,
(3) for all $(t, x) \in R_+ \times R^n_+$

$$D^+V(t, x)|_{(2.2)} \le 0.$$

Then the solutions of system (2.2) are (ρ, ρ_0)-bounded.

See [4] for a proof of Theorem 3.1.

Let us give a criterion for the uniform (ρ_0, ρ)-boundedness. In what follows we use notations

$$S(\rho_0, h) = \{(t, x) \in R_+ \times R^n_+ : \rho_0(t, x) < h\},$$
$$S(\rho, h) = \{(t, x) \in R_+ \times R^n_+ : \rho(t, x) < h\};$$

$S^c(\rho_0, h)$ stands for the complement of $S(\rho_0, h)$.

Theorem 3.2 *Suppose that*

(1) $\rho, \rho_0 \in \mathcal{M}$ *and* ρ *is continuous with respect to* ρ_0,
(2) $V(t,x)$ *is locally Lipschitz in* x, $a \in \mathcal{KR}$, $w \in C(R_+, R_+)$, *and*

$$a(\rho(t,x)) \leq V(t,x) \leq w(\rho_0(t,x))$$

for all $(t,x) \in S^c(\rho_0, h)$ *with some* $h > 0$,
(3) for all $(t,x) \in S^c(\rho_0, h)$

$$D^+V(t,x)|_{(2.2)} \leq 0.$$

Then the solutions of system (2.2) are uniformly (ρ, ρ_0)*-bounded.*

Proof For every $\alpha > 0$ choose $\gamma > 0$ so that

$$a(\gamma) > \max\{w(\max\{\alpha, h\}), a^{-1}(\phi(h))\}. \tag{3.2}$$

Here ϕ is a function introduced in Definition 3.4, (ii). Since by Definition 3.3 $\lim_{r\to\infty} a(r) = \infty$, there is γ satisfying (3.2). Let $t_0 \in R_+$ and $\rho_0(t_0, x_0) < \alpha$. Assume that for some solution of system (2.2), $x(t)$, there is $t^* \geq t_0$ such that

$$\rho(t^*, x(t^*)) \geq \gamma.$$

By (3.2) $\gamma \geq \phi(h)$, hence by the definition of ϕ, we have $\rho_0(t^*, x(t^*)) > h$. Then by the continuity of $x(t)$ there exist t_1 and t_2, $t_0 \leq t_1 < t_2 \leq t^*$, such that

$$\rho_0(t_1, x(t_1)) = \max\{\alpha, h\},$$

$$(t, x(t)) \in S(\rho, \gamma) \cap S^c(\rho_0, \max\{\alpha, h\}) \quad \text{for all} \quad t \in [t_1, t_2].$$

By condition (2)

$$V(t_1, x(t_1)) \leq w(\rho_0(t_1, x(t_1)) = w(\max\{\alpha, h\}) \tag{3.3}$$

and

$$a(\gamma) = a(\rho(t_2, x(t_2))) \leq V(t_2, x(t_2)). \tag{3.4}$$

By condition (3)

$$V(t_2, x(t_2)) \leq V(t_1, x(t_1)). \tag{3.5}$$

Estimates (3.3)–(3.5) imply that

$$a(\gamma) \leq w(\max\{\alpha, h\})$$

which contradicts (3.2). Therefore $\rho(t, x(t)) < \gamma$ for all $t \geq t_0$ whenever $\rho_0(t_0, x_0) \leq \alpha$, i.e., the solutions of system (2.2) are uniformly (ρ_0, ρ)-bounded.

3.2 *Perturbed Kolmogorov System*

For the perturbed system (2.1) we shall utilize $m \times m$ matrix functions

$$U_1(t,x) = [u_{1ij}(t,x)], \quad U_2(t,x,\mu) = [u_{2ij}(t,x,\mu)], \quad i, j = 1, 2, \ldots, m, \tag{3.6}$$

where $u_{1ij} \in C(R_+ \times R^n_+, R)$, $u_{2ij} \in C(R_+ \times R^n_+ \times M^k, R)$. Take $\eta \in R^m$ and set

$$V_1(t,x) = \eta^T U_1(t,x)\eta, \quad V_2(t,x,\mu) = \eta^T U_2(t,x,\mu)\eta.$$

Define the derivatives of V_1 and V_2 along the vector field (2.1) at point (t,x) by

$$D^+V_1(t,x)|_{(2.1)} = \limsup_{\theta \to 0+}[V(t+\theta, x+\theta \operatorname{diag}(\beta(x))F(t,x,\mu)) - V_1(t,x)]\theta^{-1}.$$

$$D^+V_2(t,x,\mu)|_{(2.1)} = \limsup_{\theta \to 0+}[V_2(t+\theta, x+\theta \operatorname{diag}(\beta(x))F(t,x,\mu)) - V(t,x,\mu)]\theta^{-1}.$$

Here $\operatorname{diag}(\beta(x))$ is a diagonal $n \times n$ matrix with $\beta_1(x_1), \ldots, \beta_n(x_n)$ on the diagonal.

Theorem 3.3 *Suppose that*

(1) $\rho, \rho_0 \in \mathcal{M}$ and ρ is continuous with respect to ρ_0,
(2) $V_1(t,x)$ is locally Lipschitz in x,

$$a(\rho(t,x)) \leq V_1(t,x) \leq w(t, \rho_0(t,x))$$

for all $(t,x) \in R_+ \times R^n_+$, where $a \in \mathcal{KR}$, $w \in C(R^2_+, R_+)$, and

$$D^+V_1(t,x)|_{(2.2)} \leq g_1(t, V_1(t,x))$$

for all $(t,x) \in R_+ \times R^n_+$, where $g_1 \in C(R^2_+, R)$ and $g_1(t,u)$ is nondecreasing in u for all $t > t'_0$,
(3) for all $(t,x) \in S^c(\rho_0, h)$ with some $h > 0$ and all $\mu \in M^k$

$$|V_2(t,x,\mu)| < c(\mu),$$

where $c \in C(M^k, R_+)$, $c(0) = 0$, and

$$D^+V_2(t,x)|_{(2.1)} + D^+V_2(t,x,\mu)|_{(2.1)} \leq g_2(t, V_1(t,x) + V_2(t,x,\mu), \mu)$$

where $g_2 \in C(R^2_+ \times M^k, R)$ and $g_2(t,u,\mu)$ is nondecreasing in u for all $t > t'_0$,
(4) every solution of the equation

$$\frac{du}{dt} = g_1(t,u), \quad u(t_0) = u_0 \geq 0$$

is bounded,
(5) the solutions of the equations

$$\frac{dw}{dt} = g_2(t,w,\mu), \quad w(t_0) = w_0 \geq 0$$

are bounded uniformly in $\mu \in M^k$.

Then the solutions of system (2.1) are (ρ, ρ_0, μ)-bounded.

For a proof we refer to [4].

4 Stability with Respect to two Measures

4.1 Unperturbed Kolmogorov System

Definition 4.1 System (2.2) is called (ρ_0, ρ)-stable if there exists a nonnegative function $\delta(t_0, \epsilon)$ on R_+^2 continuous in t_0, satisfying $\delta(t_0, \epsilon) > 0$ for $\epsilon > 0$, and such that for every $\epsilon > 0$ and every $(t_0, x_0) \in R_+ \times R_+^n$, solution $x(t)$ of (2.2), satisfies

$$\rho(t, x(t)) < \epsilon \quad \text{for all} \quad t \geq t_0$$

whenever $\rho_0(t_0, x_0) < \delta(t_0, \epsilon)$.

Again, we use the Lyapunov function V (3.1).

Theorem 4.1 *Suppose that*

(1) $\rho, \rho_0 \in \mathcal{M}$ *and* ρ *is continuous with respect to* ρ_0,

(2) $V(t,x)$ *is locally Lipschitz in* x, $a \in \mathcal{KR}$, $w \in C(R_+^2, R_+)$, $w(t, 0) = 0$, *and*

$$a(\rho(t,x)) \leq V(t,x) \leq w(t, \rho_0(t,x))$$

for all $(t,x) \in R_+ \times R_+^n$.

(3) for all $(t,x) \in S(\rho, h)$ *with some* $h > 0$ *it holds that*

$$D^+V(t,x)|_{(2.2)} \leq 0.$$

Then the solutions system (2.2) is (ρ, ρ_0)*-stable.*

This theorem follows from Theorem 4.1 in [3].

4.2 Perturbed Kolmogorov system

Definition 4.2 System (2.1) is called (ρ_0, ρ, μ)-stable if there exist a nonnegative function $\delta(t_0, \epsilon)$ on R_+^2 continuous in t_0 and satisfying $\delta(t_0, \epsilon) > 0$ for $\epsilon > 0$, and a function $\mu^*(t_0, \epsilon) : R_+^2 \mapsto M^k \setminus \{0\}$ such that for every $\epsilon > 0$ every $(t_0, x_0) \in R_+ \times R_+^n$, and every $\mu \in M^k$, solution $x(t, \mu)$ of (2.1) satisfies

$$\rho(t, x(t, \mu)) < \epsilon \quad \text{for all} \quad t \geq t_0$$

whenever $\rho_0(t_0, x_0) < \delta(t_0, \epsilon)$ and $\mu_i < \mu_i^*(t_0, \epsilon)$, $i = 1, \ldots, k$.

Below, we refer to the Lyapunov functions V_1 and V_2 (3.6).

Theorem 4.2 *Suppose that*

(1) $\rho, \rho_0 \in \mathcal{M}$ *and* ρ *is continuous with respect to* ρ_0,

(2) $V_1(t,x)$ *is locally Lipschitz in* x, *for all* $(t,x) \in R_+ \times R_+^n$ *it holds that*

$$a(\rho(t,x)) \leq V_1(t,x) \leq w(t, \rho_0(t,x)),$$

where $a \in \mathcal{K}$, $w \in C(R_+^2, R_+)$, $w(t,0) = 0$, *and for all* $(t,x) \in S(\rho,h)$ *with some* $h > 0$ *it holds that*

$$D^+ V_1(t,x)|_{(2.2)} \leq g_1(t, V_1(t,x)),$$

where $g_1 \in C(R_+^2, R)$, $g_1(t,u)$ *is nondecreasing in* u *and* $g_1(t,0) = 0$ *for* $t > t_0'$,

(3) *for all* $(t,x) \in S(\rho,h) \cup S(\rho_0,h)$ *and all* $\mu \in M^k$ *it holds that*

$$|V_2(t,x,\mu)| < c(\mu),$$

where $c \in C(M^k, R_+), c(0) = 0$, *and*

$$D^+ V_2(t,x)|_{(2.1)} + D^+ V_2(t,x,\mu)|_{(2.1)} \leq g_2(t, V_1(t,x) + V_2(t,x,\mu), \mu)$$

where $g_2 \in C(R_+^2 \times M^k, R), g_2(t,u,\mu)$ *is nondecreasing in* u *and* $g_2(t,0,\mu) = 0$ *for all* $t > t_0'$,

(4) *the zero solution of the equation*

$$\frac{du}{dt} = g_1(t,u), \quad u(t_0) = u_0 \geq 0$$

is stable,

(5) *the zero solutions of the equations*

$$\frac{dw}{dt} = g_2(t,w,\mu), \quad w(t_0) = w_0 \geq 0$$

are stable uniformly in $\mu \in M^k$.

Then system (2.1) is (ρ, ρ_0, μ)*-stable.*

A proof is similar to that of Theorem 3.3.

5 Application to a Generalized Lotka-Volterra System

Consider a generalized Lotka-Volterra equation,

$$\frac{dx}{dt} = \mathrm{diag(x)}[b + A(x)x], \quad x(t_0) = x_0, \quad i = 1,2,\ldots,n, \tag{5.1}$$

as an example of system (2.2). Here $x \in R_+^2$, $A(x) = [a_{ij}(x)]$, $i,j = 1,2$, $a_{ij} \in C(R_+^2, R), b \in R^2$. Let $x^* \in R_+^2$ be a nonzero equilibrium state of system (5.1). Assume it to be unique. Introduce Lyapunov variables

$$y_1 = x_1 - x_1^*, \quad y_2 = x_2 - x_2^*.$$

Define a matrix function

$$U(y) = [u_{ij}(y)], \quad i,j = 1,2$$

by setting

$$u_{11}(y) = \alpha y_1^2, \quad u_{12}(y) = u_{12}(y) = -\gamma y_1 y_2, \quad u_{22}(y) = \beta y_2^2$$

with $\alpha, \beta > 0$. Let $\eta \in R_+^2$ and

$$V(y) = \eta^T U(y)\eta.$$

The following estimates hold (see [3]). First, we have

$$V(y) \geq u^T(y) H^T P H u(y), \tag{5.2}$$

where $u^T(y) = (|y_1|, |y_2|)$, $H = \text{diag}(\eta)$, and

$$P = \begin{pmatrix} \alpha & -\gamma \\ -\gamma & \beta \end{pmatrix}.$$

Second, it holds that

$$D^+ V(y)|_{(5.1)} \leq u^T(y)[C(y) + G(y)]u(y), \tag{5.3}$$

where

$$C(y) = \begin{pmatrix} C_{11}(y) & C_{12}(y) \\ C_{21}(y) & C_{22}(y) \end{pmatrix}, \quad G(y) = \begin{pmatrix} \sigma_{11}(y) & \sigma_{12}(y) \\ \sigma_{21}(y) & \sigma_{22}(y) \end{pmatrix},$$

$$C_{11}(y) = 2\eta_1(\alpha\eta_1 a_{11}(y + x^*)x_1^* + \gamma\eta_2 a_{21}(y + x^*)x_2^*),$$

$$C_{22}(y) = 2\eta_2(\beta\eta_2 a_{22}(y + x^*)x_2^* + \gamma\eta_1 a_{12}(y + x^*)x_1^*),$$

$$C_{12}(y) = C_{21}(y) = \alpha\eta_1^2|a_{12}(y + x^*)|x_1^* + \beta\eta_2^2|a_{21}(y + x^*)|x_2^*$$
$$+ \eta_1\eta_2|\gamma(a_{11}(y + x^*)x_1^* + a_{21}(y + x^*)x_2^*)|,$$

and

$$\sigma_{11}(y) = 2\alpha\eta_1^2|a_{11}(y + x^*)||y_1|,$$

$$\sigma_{22}(y) = 2\beta\eta_2^2|a_{22}(y + x^*)||y_2|,$$

$$\sigma_{12}(y) = \sigma_{21}(y) =$$

$$[\alpha\eta_1^2|a_{12}(y + x^*)| + \eta_1\eta_2|\gamma||a_{21}(y + x^*) + a_{11}(y + x^*)|]|y_1| +$$

$$[\beta\eta_2^2|a_{21}(y + x^*)| + \eta_1\eta_2|\gamma||a_{22}(y + x^*) + a_{12}(y + x^*)|]|y_2|.$$

We apply Theorem 3.1 with

$$\rho(t, y) = \rho_0(t, y) = \|y\|. \tag{5.4}$$

Note that

$$\lambda_m u^T(y)u(y) \leq V(y) \leq \lambda_M u^T(y)u(y),$$

where λ_m and λ_M are, respectively, the minimum and maximum eigenvalues of matrix $H^T PH$.

Theorem 5.1 *Assume that*

(1) ρ *and* ρ_0 *are defined by (5.4),*
(2) $0 < \lambda_m < \lambda_M$,
(3) there exist 2×2 *matrixes* $\bar{C}$ *and* $\bar{G}$ *such that* $\bar{C} + \bar{G}$ *is negative semi-definite, and estimates*

$$w^T C(y) w \leq w^T \bar{C} w, \quad w^T G(y) w \leq w^T \bar{G} w$$

hold for all $y, w \in R^2_+$.

Then the solutions of system (5.1) are (ρ, ρ_0)*-bounded.*

Proof One can easily see that, in view of estimates (5.2) and (5.3) and assumptions 1–3, all conditions of Theorem 3.1 are satisfied.

The next statement follows from Theorem 4.1.

Theorem 5.2 *Assume that*

(1) conditions 1 and 2 of Theorem 5.1 are satisfied,
(2) there exist 2×2 *matrixes* C^* *and* G^* *such that* $C^* + G^*$ *is negative semi-definite, and estimates*

$$w^T C(y) w \leq w^T C^* w, \quad w^T G(y) w \leq w^T G^* w$$

hold for all $y, w \in S(\rho, h)$ *with some* $h > 0$.

Then system (5.4) is (ρ, ρ_0)-stable.

References

[1] Freedman, H.I. (1987) Deterministic mathematical models in population ecology. HIFR Consulting LDT, Edmonton.
[2] Freedman, H.I. and Martynyuk, A.A. (1993) On stability with respect to two measures in Kolmogorov's model of dynamics of populations. *Doklady Akad. Nauk Russia*, **329**, 423–425 (in Russian).
[3] Freedman, H.I. and Martynyuk, A.A. (1995) Stability analysis with respect to two measures for population growth models of Kolmogorov type. *Nonlinear Analysis*, **25**, 1221–1230.
[4] Freedman, H.I. and Martynyuk, A.A. (1995) Boundedness criteria for solutions of perturbed Kolmogorov population models. *Canadian Appl. Math. Quarterly*, **3**, 203–217.
[5] Martynyuk, A.A. (1995) Qualitative analysis of nonlinear systems by the method of matrix Lyapunov functions. *Rocky Mountain J. of Math.*, **25**, 397–415.
[6] Lakshmikantham, V., Leela, S. and Martynyuk, A.A. (1989) Stability Analysis of Nonlinear Systems. Marcel Dekker, Inc., New York.

12 GUARANTEED REGIONS OF ATTRACTION FOR DYNAMICAL POLYNOMIAL SYSTEMS

B. TIBKEN, E.P. HOFER and C. DEMIR

Department of Measurement, Control and Microtechnology, University of Ulm, D-8906 Ulm, Germany

Introduction

For the analysis of nonlinear systems given by $\dot{x} = f(x)$ it is important to be able to investigate the stability of stationary points x^* defined by $f(x^*) = 0$. Usually the solution of nonlinear state space differential equations cannot be given analytically and therefore the stability analysis and the computation of regions of attraction for asymptotically stable stationary points is complicated. In this paper the direct method of Lyapunov is used to treat these problems. Using the direct method a subset of the exact region of attraction of an asymptotically stable stationary point is calculated. For most systems this requires a nonlinear optimization. In this paper this nonlinear optimization problem is solved and a guaranteed error estimate for the solution is computed. The computed subset is the interior of the greatest equipotential surface of the Lyapunov function which lies completely in the region in which the time derivative of the Lyapunov function along the trajectories of the system is negative definite. To calculate this subset we work on an lattice, that means we calculate the derivative of the Lyapunov function on a lattice of points in the state space. Normally we cannot make any statements about the values of this function between two lattice points. Thus, the region which is calculated with a naive lattice approach may contain points with positive derivative. In this paper we solve this definiteness problem for systems with polynomial $f(x)$ and quadratic Lyapunov functions using a theorem of Ehlich and Zeller. First a small subset of the region of attraction containing the origin is guaranteed with a rough estimation. Using a special bisection method this subset is improved and inscribing and circumscribing equipotential surfaces which enclose the boundary of the subset of the region of attraction which can be guaranteed with the chosen Lyapunov function are computed. The paper concludes with the results for a typical example.

1 Statement of the Problem

In this paper nonlinear systems given by the state space representation

$$\dot{x} = f(x), \;\; x(0) = x_0 \tag{1}$$

where $x \in R^n$ is the state vector and each component of the n dimensional vector function f is a polynomial are studied. We assume $f(0) = 0$ and the asymptotic stability of $x = 0$, i.e., all eigenvalues of the Jacobian matrix of f at 0 have negative real part. The goal is to compute subsets of the region of attraction for the equilibrium $x = 0$. A region of attraction for the stationary point $x = 0$ is computed with the help of a suitably constructed Lyapunov function. For this purpose let $V(x)$ be a positive definite function near the origin, i.e., $V(0) = 0$ and $V(x) > 0$ for all $x \neq 0$ in a region around the origin. We denote the derivative of $V(x)$ along the trajectories of (1) with $\dot{V}(x)$. It is computed as $\dot{V}(x) = \{\frac{\partial V}{\partial x}(x)\}^T f(x)$ where T denotes transposition. Because of the above assumptions there exists a quadratic positive definite function $V(x)$ with a negative definite $\dot{V}(x)$ near $x = 0$. A subset Ω of the region of attraction is guaranteed as the interior of the largest level set of $V(x)$ inside the region $\{x | \dot{V}(x) < 0\}$ [1, Theorem 26.1]. This immediately leads to the optimization problem

$$c_{opt} = \min_{\dot{V}(x)=0, \; x\neq 0} V(x) \tag{2}$$

and the definition $\Omega = \{x | V(x) < c_{opt}\}$ for this subset of the exact region of attraction. The main problem is to compute the solution of the stated optimization problem (2) with guaranteed precision. In this paper a method will be presented which leads to arbitrarily close upper and lower bounds for c_{opt}. The method will be applied to a typical example.

2 Calculation of Bounds

For our approach to the estimation of subsets of the exact region of attraction we will use a result which enables us to calculate bounds for the values of a polynomial in a real interval. In the following, $J = [a, b]$ denotes a nonempty compact real interval with $J \subset R$. We define the set of Chebychev points in J for a given natural number $N > 0$ by $X(N, J) := \{x_1, x_2, \ldots, x_N\}$, where $x_i := \frac{a+b}{2} + \frac{b-a}{2}\cos\left(\frac{(2i-1)\pi}{2N}\right)$ for $i = 1, \ldots, N$. For a function h defined on a set I we define the norm $||h||^I := \max_{x\in I} |h(x)|$ which is the usual maximum norm. Let P_n be the set of polynomials in one variable with $deg P \leq n$. Then the following inequality

$$||P||^J \leq C\left(\frac{n}{N}\right) ||P||^{X(N,J)} \tag{3}$$

with

$$N > n \text{ and } C(q) := \left[\cos\left(\frac{q}{2}\pi\right)\right]^{-1} \quad \text{for} \quad 0 \leq q \leq 1 \tag{4}$$

is valid for every $P \in P_n$ and every interval J. Inequality (3) is remarkable because the norm $\|P\|^{X(N,J)}$ on the right-hand side of (3) depends on the values of P at the Chebychev points only. This result was given by Ehlich and Zeller in [2]. Using (3); the following inequalities

$$P^J_{\min} \geq \frac{1}{2}\left\{\left(C\left(\frac{n}{N}\right)+1\right)P^{X(N,J)}_{\min} - \left(C\left(\frac{n}{N}\right)-1\right)P^{X(N,J)}_{\max}\right\}, \tag{5}$$

$$P^J_{max} \leq \frac{1}{2}\left\{\left(C\left(\frac{n}{N}\right)+1\right)P^{X(N,J)}_{\max} - \left(C\left(\frac{n}{N}\right)-1\right)P^{X(N,J)}_{\min}\right\} \tag{6}$$

which are valid for every $P \in P_n$ and $N > n$ are given by Gärtel in [3] where $P^I_{min} := \min_{x \in I} P(x)$ and $P^I_{max} := \max_{x \in I} P(x)$ are the maximum and minimum of P in the set I, respectively. For trigonometric polynomials and for rational functions similar inequalities are given by Gärtel. We will give these results at the end of this section.

The inequalities (3, 5, 6) are valid for polynomials in one variable; they are extended to polynomials of several variables using the following replacements. The interval J is replaced by

$$\hat{J} = [a_1, b_1] \times [a_2, b_2] \times \cdots \times [a_n, b_n] \tag{7}$$

which represents a hyperrectangle. For the degree of P with respect to the i-th variable x_i we introduce the abbreviation n_i and the set of Chebychev points in $\hat{J}$ is given by

$$X(\hat{N}, \hat{J}) := X(N_1, [a_1, b_1]) \times \cdots \times X(N_n, [a_n, b_n]) \tag{8}$$

where N_i is the number of Chebychev points in the interval $[a_i, b_i]$. Then the inequalities

$$P^{\hat{J}}_{\min} \geq \frac{1}{2}\left\{(K+1)P^{X(\hat{N},\hat{J})}_{\min} - (K-1)P^{X(\hat{N},\hat{J})}_{\max}\right\}, \tag{9}$$

$$P^{\hat{J}}_{\max} \leq \frac{1}{2}\left\{(K+1)P^{X(\hat{N},\hat{J})}_{\max} - (K-1)P^{X(\hat{N},\hat{J})}_{\min}\right\} \tag{10}$$

with

$$K = \prod_{i=1}^{n} C\left(\frac{n_i}{N_i}\right) \tag{11}$$

under the conditions $N_i > n_i$, $i = 1, \ldots, n$, are valid. The given inequalities are now extended to trigonometric polynomials. A trigonometric polynomial of degree n in the variable φ is defined as

$$P(\varphi) := a + \sum_{k=1}^{n}(\alpha_k \sin(k\varphi) + \beta_k \cos(k\varphi)) \tag{12}$$

which is just a finite Fourier series. Let P be a trigonometric polynomial with $degP \leq n$. Then the following inequality

$$\|P\|^{[0,2\pi]} \leq C\Big(\frac{n}{N}\Big)\|P\|^{\Psi(N)} \tag{13}$$

with

$$\Psi(N) = \left\{\frac{(i-1)\pi}{N}, i = 1, \ldots, 2N\right\} \tag{14}$$

and $N > n$ is valid. The inequalities (5, 6) are generalized to

$$P_{min}^{[0,2\pi]} \geq \frac{1}{2}\Big\{(K+1)P_{min}^{\Psi(N)} - (K-1)P_{max}^{\Psi(N)}\Big\}, \tag{15}$$

$$P_{max}^{[0,2\pi]} \leq \frac{1}{2}\Big\{(K+1)P_{max}^{\Psi(N)} - (K-1)P_{min}^{\Psi(N)}\Big\} \tag{16}$$

with $K = C(\frac{n}{N})$. The generalization to several variables leads to (10, 11) with the replacements $\hat{J} = [0, 2\pi]^n$ and $X(\hat{N}, \hat{J}) = \Psi(N_1) \times \Psi(N_2) \times \cdots \times \Psi(N_n) = \Psi(\hat{N})$ and with K given by (11). Exactly the same results hold if the function under investigation is a polynomial in some of the variables and a trigonometric polynomial in the remaining variables. That means if $P(x_1, x_2, \ldots, x_n)$ is a trigonometric polynomial in the variables $x_1, \ldots, x_r$ and a polynomial in the variables $x_{r+1}, \ldots, x_n$ with the corresponding degrees $n_1, \ldots n_n$, the bounds (9,10) are valid with $\hat{J} = [0, 2\pi]^r \times [a_{r+1}, b_{r+1}] \times \cdots [a_n, b_n]$, $X(\hat{N}, \hat{J}) = \Psi(N_1) \times \cdots \times \Psi(N_r) \times X(N_{r+1}, [a_{r+1}, b_{r+1}]) \times \cdots \times X(N_n, [a_n, b_n])$ and K given by (11). We apply these bounds in the next sections to compute upper bounds on polynomials and trigonometric polynomials in several variables based on values at a finite number of points.

3 Solution of the Optimization Problem

In this section we describe the application of the theorems of Ehlich and Zeller and of Gärtel to the computation of regions of attraction. Quadratic Lyapunov functions given by

$$V(x) = x^T Q x \tag{17}$$

where Q is a positive definite symmetric matrix will be used to compute subsets of the region of attraction as ellipsoids. The time derivative of $V(x)$ along the trajectories of (1) is calculated as $\dot{V}(x) = 2f^T(x)Qx$. Due to the assumption $f(0) = 0$ and the property that all eigenvalues of $A = \frac{\partial f}{\partial x}(0)$ have strictly negative real parts, the matrix Q is chosen as a solution of the Lyapunov equation

$$A^T Q + QA = -P \tag{18}$$

with $P = P^T > 0$. In the following we assume that a specific Q has been calculated. The level sets $\Omega_x = \{y | V(y) < V(x)\}$ represent the interior of ellipsoids in R^n. Our goal is the calculation of the largest level set Ω^* such that $\dot{V}(x)$ is negative definite in Ω^*. According to the first section we have $\Omega^* = \{x | V(x) < c_{opt}\}$ where c_{opt} is given by (2). Here we will use the results of the previous section to compute upper and lower bounds for c_{opt}. First we compute the Cholesky decomposition $Q = L^T L$ of the matrix Q where L is a regular upper triangular matrix [4]. Using this L we change variables according to $y = Lx$ which leads to the representations

$$\tilde{V}(y) = V(L^{-1}y) = y^T y \tag{19}$$

and

$$\dot{\tilde{V}}(y) = \dot{V}(L^{-1}y) \tag{20}$$

respectively. The crucial fact is that Ω_x is transformed into the set

$$\tilde{\Omega}_x = \{z | z^T z < V(x)\} \tag{21}$$

which is a hypersphere. Thus, we have to look for the largest hypersphere such that $\dot{\tilde{V}}(y)$ is negative in the interior of this sphere except at the origin where it is zero. To start our calculation we assume that we already have computed an r_0 with the property that $\dot{\tilde{V}}(y)$ is negative definite in $\{y \,|\, ||y|| \leq r_0\}$. A method to compute such an r_0 is described in the next section. Now the idea of the method described in detail below is to increase r_0 thus; we choose $r = 2r_0$ and we can guarantee that $\dot{\tilde{V}}(x)$ is negative definite in the larger sphere $\{y \,|\, ||y|| \leq r\}$ if it is negative in $\Delta = \{y \,|\, r_0^2 \leq y^T y \leq r^2\}$ which is the region between the two hyperspheres with radius r_0 and r, respectively. To verify negativity in Δ the n-dimensional polar coordinates $\rho, \alpha_1, \ldots, \alpha_{n-1}$ are introduced by

$$\begin{aligned} y_1 &= \rho \cos(\alpha_1) \cdots \cos(\alpha_{n-3}) \cos(\alpha_{n-2}) \cos(\alpha_{n-1}), \\ y_2 &= \rho \cos(\alpha_1) \cdots \cos(\alpha_{n-3}) \cos(\alpha_{n-2}) \sin(\alpha_{n-1}), \\ y_3 &= \rho \cos(\alpha_1) \cdots \cos(\alpha_{n-3}) \sin(\alpha_{n-2}), \\ &\vdots \\ y_n &= \rho \sin(\alpha_1). \end{aligned} \tag{22}$$

Using these coordinates the set Δ is the image of $[r_0, r] \times [0, 2\pi] \times [0, \pi] \times \cdots \times [0, \pi]$ which means that $\dot{\tilde{V}}(y)$ is a polynomial in ρ and a trigonometric polynomial in $(\alpha_1, \ldots, \alpha_{n-1})$. Now an upper bound for $\dot{\tilde{V}}(y)$ in Δ is computed using the results of the previous section. If, during the computation of $\dot{\tilde{V}}(y)$ at the lattice points, a positive value is found the whole computation is interrupted and the radius $\tilde{r}$ of this point is an upper bound, for the radius of the sphere lying entirely in $\dot{\tilde{V}}(y) < 0$. If the upper bound of $\dot{\tilde{V}}(y)$ in Δ is negative we set $r_0 = r$, choose again $r = 2r_0$ and repeat the process. If the upper bound is positive but all function values on the lattice are negative, the number of points in the lattice is increased and the calculation is repeated until $\dot{\tilde{V}}(y) < 0$ for all $y \in \Delta$ is guaranteed or a point in Δ with a positive value for $\dot{\tilde{V}}(y)$ is found. If the region of attraction is not the whole of R^n after a finite number of steps, an upper bound for the radius is found. When upper and lower bounds for the radius are determined, both bounds are improved by bisection. The algorithm is given by the following steps:

1. Compute r_0 with $\dot{\tilde{V}}(x) < 0$ for all x with $x^T x \leq r_0^2$ and $x \neq 0$; this is done with the techniques described in the next section. Set $r_l = r_0, r_u = \infty$ and go to step 2.
2. If $r_u - r_l$ is less than a given tolerance ϵ go to step 6, else go to step 3.
3. If $r_u = \infty$, set $r = 2r_l$, else set $r = \frac{1}{2}(r_l + r_u)$. Go to step 4.
4. Compute $\dot{\tilde{V}}(x)$ at the lattice points in $[r_l, r] \times [0, 2\pi] \times \cdots \times [0, \pi]$ introduced in Section 2. Go to step 5.
5. If $\dot{\tilde{V}}(x)$ is positive at a lattice point set r_u to the corresponding radius and go to step 2. If $\dot{\tilde{V}}(x)$ can be ensured to be negative in Δ by the bounds of Section 2 set $r_l = r$ and go to step 2. If the bound is positive but all function values of $\dot{\tilde{V}}(x)$ at the lattice points are negative increase the number of lattice points and repeat step 4.
6. Stop.

This algorithm converges in a finite number of steps to bounds r_l and r_u for the optimal radius which satisfy $r_u - r_l < \epsilon$ for any given $\epsilon > 0$ which measures the accuracy of the computed solution. After the algorithm stops it is guaranteed that $\{y \,|\, y^T y \leq r_l^2\}$ is completely inside the region with $\dot{\tilde{V}}(y) < 0$ and the sphere $\{y \,|\, y^T y \leq r_u^2\}$ contains at least one point with $\dot{\tilde{V}}(x) > 0$. Thus, due to (21) the sphere $\Omega = \{x \,|\, V(x) \leq r_l^2\}$ is a subset of the region of attraction for the stationary solution $x = 0$ of (1) which is guaranteed with certainty. This is the largest subset of the region of attraction which can be guaranteed with the chosen Lyapunov function.

4 Initial Estimate

In this section a sphere around the origin in which the polynomial $\dot{\tilde{V}}(x)$ is negative definite is computed. For the following computations the function $\dot{\tilde{V}}(x)$ is represented in the form

$$\dot{\tilde{V}}(x) = \dot{\tilde{V}}_2(x) + \dot{\tilde{V}}_3(x) + \cdots + \dot{\tilde{V}}_m(x) \tag{23}$$

where $\dot{\tilde{V}}_i(x)$ represents all terms of degree i in the polynomial $\dot{\tilde{V}}(x)$. Here the fact has been used that there are no constant and linear terms in the expansion of $\dot{\tilde{V}}(x)$. Each term $\dot{\tilde{V}}_i(x)$ is written as

$$\dot{\tilde{V}}_2(x) = \sum_{i_1 + \cdots + i_n = 2} a_{i_1 \cdots i_n} x_1^{i_1} \cdots x_n^{i_n}$$

$$= x^T \tilde{P} x, \text{where } \tilde{P} \text{ is a real symmetric } n \times n - matrix,$$

$$\dot{\tilde{V}}_3(x) = \sum_{i_1+\cdots+i_n=3} a_{i_1\cdots i_n} x_1^{i_1} \cdots x_n^{i_n},$$

$$\dot{\tilde{V}}_4(x) = \sum_{i_1+\cdots+i_n=4} a_{i_1\cdots i_n} x_1^{i_1} \cdots x_n^{i_n},$$

$$\vdots$$

$$\dot{\tilde{V}}_m(x) = \sum_{i_1+\cdots+i_n=m} a_{i_1\cdots i_n} x_1^{i_1} \cdots x_n^{i_n}.$$

Now the following bound is established for all $\dot{\tilde{V}}_k(x)$ with $k \geq 3$

$$\begin{aligned} \max_{x^T x \leq r^2} |\dot{\tilde{V}}_k(x)| &\leq \sum_{i_1+\cdots+i_n=k} |a_{i_1\cdots i_n}| \max_{x^T x \leq r^2} (|x_n|^{i_1} \cdots |x_n|^{i_n}), \\ &= \sum_{i_1+\cdots+i_n=k} |a_{i_1\cdots i_n}| r^k = c_k r^k \end{aligned} \tag{25}$$

where the coefficients

$$c_k = \sum_{i_1+\cdots+i_n=k} |a_{i_1\cdots i_n}| \tag{26}$$

have been introduced.

The quadratic term $\dot{\tilde{V}}_2(x)$ is bounded using

$$\lambda_{min}(\tilde{P}) x^T x \leq \dot{V}_2(x) \leq \lambda_{max}(\tilde{P}) x^T x = \lambda_{max}(\tilde{P}) r^2 \tag{27}$$

where $\tilde{P}$ is a negative definite matrix given by $-L^{-1^T} P L^{-1}$ with P introduced in (18) and L given by the Cholesky decomposition of Q. For $\dot{\tilde{V}}(x)$ we compute

$$\begin{aligned} \dot{\tilde{V}}(x) &= \dot{\tilde{V}}_2 + \dot{\tilde{V}}_3 + \cdots + \dot{\tilde{V}}_m \leq \dot{\tilde{V}}_2 + |\dot{\tilde{V}}_3| + \cdots + |\dot{\tilde{V}}_m|, \\ &\leq \lambda_{max}(\tilde{P}) r^2 + \max |\dot{\tilde{V}}_3| + \cdots + \max |\dot{\tilde{V}}_m|, \\ &\leq \lambda_{max}(\tilde{P}) r^2 + c_3 r^3 + \cdots + c_m r^m \end{aligned} \tag{28}$$

where $r^2 = x^T x$ has been used. This estimate is used to calculate a sphere around the origin in which $\dot{\tilde{V}}(x)$ is guaranteed to be negative. Rewriting (28) as

$$\dot{\tilde{V}}(x) \leq r^2 [\lambda_{max}(\tilde{P}) + c_3 r + c_4 r^2 + \cdots + c_m r^{m-2}] = r^2 p(r) \tag{29}$$

with $p(r) = \lambda_{max}(P) + c_3 r + c_4 r^2 + \cdots + c_m r^{m-2}$ it is clear that $\dot{\tilde{V}}$ is negative in a sphere with radius r^* if $p(r^*)$ is negative. It is easy to compute $p(0) = \lambda_{max}(\tilde{P}) < 0$ and thus the radius of the largest sphere which can be guaranteed from the estimate (28) is given by the smallest real positive root of $p(r)$. Because all coefficients of $p(r)$ are positive except the first the polynomial $p(r)$ is strictly monotone increasing for real positive arguments and thus has exactly one positive real root. This positive real root is used as r_0 in the algorithm presented in the previous section.

5 Example

In this section the presented algorithm is applied to a typical example. The state space representation of the system to be investigated is given by

$$\dot{x}_1 = f_1 = -x_1 + 2x_1^3 x_2, \quad (30)$$

$$\dot{x}_2 = f_2 = -x_2 \quad (31)$$

which is a polynomial system. Here f_1 is a polynomial of degree 3 with respect to x_1 and degree 1 with respect to x_2. The polynomial f_2 is linear. As a Lyapunov function $V(x) = x_1^2 + x_2^2$ is chosen. For this Lyapunov function the exact value of c_{opt} is $c_{opt} = \frac{5}{8}\sqrt[3]{\frac{25}{2}} = 1.450\ldots$. The lower bound $1.44 < c_{opt}$ is computed with the algorithm presented in this paper. In Figure 1 the sets $\{x \mid \dot{V}(x) = 0\}$ and $\{x \mid V(x) = c_{opt}\}$ are plotted. It is observed that the optimization problem has two different optimal points, namely, the points where the two sets intersect. Thus, the optimization problem (2) has two optimal points corresponding to the global minimum.

As a second application the same system with a different Lyapunov function is investigated. The results are shown in Figure 2. In this case it is important to observe that the optimization problem (2) has only one optimal point for the global minimum but from the subset of $\{x \mid \dot{V}(x) = 0\}$ which lies in the first quadrant it is

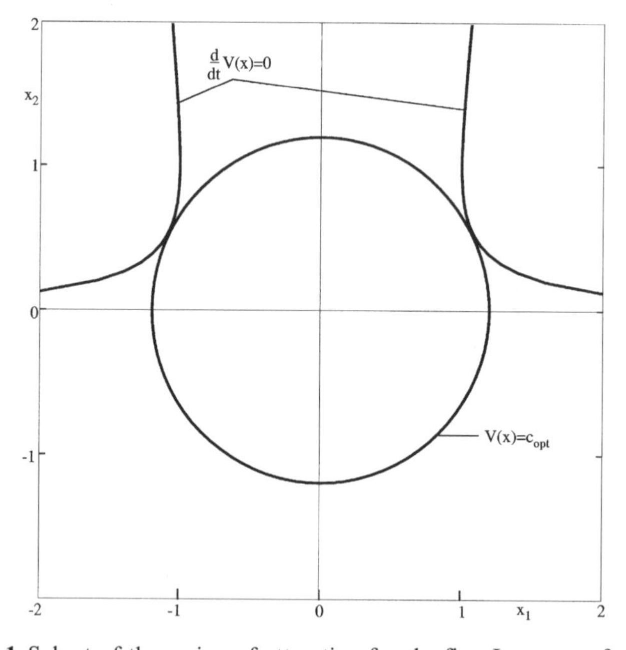

Figure 1 Subset of the region of attraction for the first Lyapunov function.

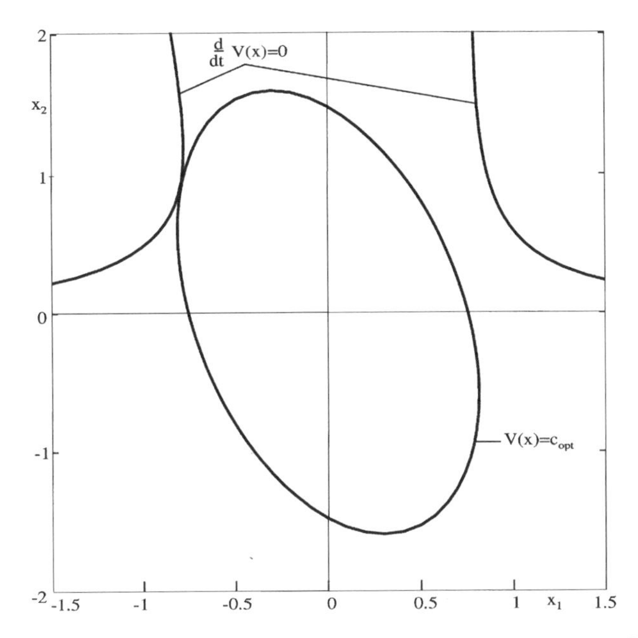

Figure 2 Subset of the region of attraction for the second Lyapunov function.

observed that at least one local minimum exists in this region of state space. This will lead to severe problems with traditional optimization algorithms because the convergence to the global minimum usually cannot be guaranteed. The algorithm presented here has no problem in computing the global solution to the optimization problem (2) even in this case. This problem of local minima is usually even worse in higher dimensional problems.

6 Conclusions and Outlook

In this paper asymptotically stable stationary points of polynomial systems have been investigated. A new algorithm for the computation of a subset of the region of attraction based on a quadratic Lyapunov function has been presented. With this algorithm the solution of the corresponding global optimization problem can be computed to any given accuracy. An example presented in this paper illustrates the results which can be achieved with this new algorithm. The work presented here has been extended to Lyapunov functions which are polynomials of degree 3 in the state variables. This extension will be published in the near future. The method can be extended to systems where the right-hand sides of the state space differential equations are continuously differentiable because these systems can be approximated arbitrarily well by a polynomial system. This extension will be investigated in future work.

References

[1] Hahn, W. (1967) *Stability of Motion*. Springer, Berlin.
[2] Ehlich, H. and Zeller, K. (1964) Schwankung von Polynomen zwischen Gitterpunkten. *Math. Z.*, **86**, 41–44.
[3] Gärtel, U. (1987) Fehlerabschätzungen für vektorwertige Randwertaufgaben zweiter Ordnung, insbesondere für Probleme aus der chemischen Reaktions-Diffusions-Theorie. Ph.D. Thesis, University of Cologne.
[4] Stoer, J. (1989) Numerische Mathematik 1, 5. Auflage, Springer, Berlin.

13 DYNAMICS FOR BIMATRIX GAMES VIA ANALYTIC CENTERS

ARKADII V. KRYAZHIMSKII[1] and GYÖRGY SONNEVEND[2]

[1] *Mathematical Steklov Institute, Russian Academy of Sciences, Moscow 117966, Russia*
[2] *Department of Numerical Analysis, Eötvös University, H-1088 Budapest, Hungary*

The method of analytic centers known in convex programming is implemented for the construction of paths leading to equilibrium points in mixed strategy bimatrix games. A bimatrix game is extended to a family of time-parametrized perturbed games in which the payoffs are logarithmically penalized for the approach to the boundary of the strategy space. In the interior of the strategy space the penalties' relative weights vanish as time goes to infinity. It is shown that the Nash equilibria in the perturbed games converge to those in the unperturbed game. Moreover, equilibrium paths starting in a connected set converge to a same equilibrium, and, under appropriate nondegeneracy conditions, "almost all" equilibrium paths converge to a single interior equilibrium.

1 Introduction

Homotopy, or path following, optimization methods (see, e.g., [1]) rest on the idea of approaching an optimum along the family of solutions of time-parametrized perturbed optimization problems. The initial perturbed problem is normally chosen simple enough so that the initial point on a solution path is easily identified, and the final problem is the unperturbed one. The key questions are concerned with the justification of the existence of a (smooth) solution path, and the description of the path following dynamics. We refer to the methods of analytic centers for convex programming (see [2], [3]) as a particular field in which the homotopy idea has been successfully implemented.

In games theory the homotopy idea has not been frequently used. In [1] a general path following methodology for finding equilibria — without a detailed analysis of the existence of the solution paths — was presented. In [4] a homotopy method was in the base of a proof of the oddness of the number of equilibrium points in a bimatrix game.

The homotopy method of [4] is an obvious game-theoretic counterpart of the method of analytic centers (central paths) for linear programming. This observation is a starting point in our analysis. In the present paper we explore some further properties of the central path dynamics in bimatrix games.

A radical difference between the optimizers in extremal problems and equilibria in games is that all optimizers provide the same payoff (objective value), whereas different equilibria differ in payoffs. There is no universal criterion for the identification of a "best" equilibrium. Approaches to the selection of equilibria were studied in [5]. An equilibrium selection game dynamics inspired by the method of fictitious play, [6], was proposed in [7].

Our main result states that, under appropriate nondegeneracy conditions, all central path trajectories — apart of those originating from a small neighborhood of the boundary of the strategy space — select a single interior equilibrium, i.e., converge to this equilibrium as time goes to infinity. The interior equilibrium has two exceptional features easily treatable as advantages. First, this equilibrium is, generically, unique, and, second, represents all pure strategies with nonzero proportions. From a game-evolutionary point of view (see [8]) the latter property is interpreted as nonextinction of all phenotypes; in this sense the interior equilibrium is a favorable starting point for further evolution.

2 Central Path Equation

Consider a bimatrix game with $m \times n$ payoff matrixes A^0 and B^0 ($n, m \geq 2$). Mixed strategies of the players 1 and 2 are as usual identified with points in S_n^0 and S_m^0, respectively. Here $S_k^0 = \{z : z_1 + \cdots + z_k = 1,\ z_1, \ldots, z_k \geq 0\}$; z_i stands for the ith coordinate of a vector z; finite-dimensional vectors are treated as columns. A mixed strategy (Nash) equilibrium, $(\hat{x}^0, \hat{y}^0)$, is defined by $\hat{x}^0 \in \mathrm{Argmax}\{p_{A^0}(x^0, \hat{y}^0) : x^0 \in S_n^0\}$, $\hat{y}^0 \in \mathrm{Argmax}\{p_{B^0}(\hat{x}^0, y^0) : y^0 \in S_m^0\}$, where $p_{A^0}(x^0, y^0) = y^{0T} A^0 x^0$ and $p_{B^0}(x^0, y^0) = y^{0T} B^0 x^0$ are the payoffs to players 1 and 2, respectively, at a mixed strategy pair (x^0, y^0). We write $\mathrm{Argmax}\{r(z) : z \in E\}$ for the set of all maximizers of a scalar function $r(\cdot)$ on a set E; a unique maximizer is denoted $\mathrm{argmax}\{r(z) : z \in E\}$.

Introduce a family of perturbed games parametrized by a nonnegative time parameter t. The perturbed game at time t is called briefly the *t-perturbed* game. In the t-perturbed game players' strategy spaces are again S_n^0 and S_m^0, and the payoffs to players 1 and 2 are given by

$$p_{A^0}(t, x^0, y^0) = t p_{A^0}(x^0, y^0) + \sum_{i=1}^{n} \log x_i^0 + \phi_1^T x^0,$$

$$p_{B^0}(t, x^0, y^0) = t p_{B^0}(x^0, y^0) + \sum_{j=1}^{m} \log y_j^0 + \phi_1^T x^0,$$

where ϕ_1 and ϕ_2 are fixed vectors in R^n and R^m, respectively. The logarithmic terms penalize for the approach to the relative boundary of the mixed strategy space $S_n^0 \times S_m^0$; as some pure strategy fractions x_i^0, y_j^0 go to zero, the penalty terms go to $-\infty$. An *equilibrium* in the t-perturbed game, $(x^0(t), y^0(t))$, is defined by $x^0(t) = \mathrm{argmax}\{p_{A^0}(t, x^0, y^0(t)) : x^0 \in \mathrm{int}\ S_n^0\}$, $y^0(t) = \mathrm{argmax}\{p_{B^0}(t, x^0(t), y^0) : y^0 \in \mathrm{int}\ S_m^0\}$; here int S_k^0 stands for the relative interior of S_k^0, i.e., the set of all $z \in S_k^0$ such that $z_i > 0$ $(i = 1, \ldots, k)$. The usage of argmax instead of Argmax is correct since the maps $x^0 \mapsto p_{A^0}(t, x^0, y^0)$ and $y^0 \mapsto p_{B^0}(t, x^0, y^0)$ are strictly concave.

Lemma 1 *For every $t \geq 0$ there exists an equilibrium in the t-perturbed game.*

Proof The function $P: (x^0, y^0) \mapsto (\mathrm{argmax}\{p_{A^0}(t, u^0, y^0) : u^0 \in \mathrm{int}\ S_n^0\}, \mathrm{argmax}\{p_{A^0}(t, x^0, v^0) : v^0 \in \mathrm{int}\ S_m^0\})$ defined on int $S_m^0 \times$ int S_n^0 is continuous and obviously takes values in certain convex compactum $K \subset S_m^0 \times S_n^0$ strictly separated from the relative boundary of $S_m^0 \times S_n^0$. By Brower's theorem P has a fixed point in K. Every fixed point of P is obviously an equilibrium in the t-perturbed game.

Mixed strategies x^0 and y^0 of players 1 and 2 are uniquely determined by the vectors composed of their first $n-1$ and $m-1$ coordinates, respectively. Thus from now on we treat mixed strategies of players 1 and 2 as elements of S_{n-1} and S_{m-1}, respectively, where $S_k = \{z : z_1 + \cdots + z_k \leq 1,\ z_1, \ldots, z_k \geq 0\}$. We set $S = S_{n-1} \times S_{m-1}$. Using partitions

$$A^0 = \begin{pmatrix} A^{00} & b_1 \\ c_1^T & d_1 \end{pmatrix}, \qquad B^0 = \begin{pmatrix} B^{00} & b_2 \\ c_2^T & d_2 \end{pmatrix},$$

where A^{00} and B^{00} are $(m-1) \times (n-1)$ matricies, $b_1, b_2 \in R^{n-1}$, $c_1, c_2 \in R^{m-1}$, $d_1, d_2 \in R^1$, we represent the payoffs to players 1 and 2 at a mixed strategy pair $(x, y) \in S$ as

$$p_A(x, y) = y^T A x + g_1^T x + h_1^T y + d_1,$$
$$p_B(x, y) = y^T B x + g_2^T x + h_2^T y + d_2;$$

here $A = A^{00} - C_1 - B_1 + D_1$, $g_1 = c_1 - \bar{d}_1$, $h_1 = b_1 - \bar{d}_1$, $B = B^{00} - C_2 - B_2 + D_2$, $g_2 = c_2 - \bar{d}_2$, $h_2 = b_2 - \bar{d}_2$, $B_k = (b_k, \ldots, b_k)$, $C_k^T = (c_k, \ldots, c_k)$, $\bar{d}_k^T = (d_k, \ldots, d_k)$ $(k = 1, 2)$. The equilibria in the unperturbed bimatrix game are defined by $\bar{x} \in \mathrm{Argmax}\{p_A(x, \bar{y}) : x \in S_{n-1}\}$, $\bar{y} \in \mathrm{Argmax}\{p_B(\bar{x}, y) : y \in S_{m-1}\}$. The payoffs in the t-perturbed game take the form

$$p_A(t, x, y) = t p_A(x, y) + \sum_{i=1}^{n} \log x_i + \log\left(1 - \sum_{i=1}^{n} x_i\right) + \psi_1^T x + \phi_{1n},$$

$$p_B(t, x, y) = t p_B(x, y) + \sum_{i=j}^{m} \log y_j + \log\left(1 - \sum_{j=1}^{n} y_j\right) + \psi_1^T y + \phi_{2n},$$

where $\psi_{1i} = \phi_{1i} - \phi_{1n}$ $(i = 1, \ldots, n-1)$, $\psi_{2i} = \phi_{2i} - \phi_{2m}$ $(i = 1, \ldots, m-1)$, and an equilibrium in the t-perturbed game, $(x(t), y(t))$, is given by $x(t) = \mathrm{argmax}\{p_A(x,$

$y(t)) : x \in S_{n-1}\}$, $y(t) = \text{argmax}\{p_B(x(t), y) : y \in S_{m-1}\}$. By Lemma 1 for every $t \geq 0$ there exists an equilibrium in the t-perturbed game; moreover, it lies in the interior of S further denoted by int S. Due to the strong concavity of $x \mapsto p_A(t, x, y)$ and $y \mapsto p_B(t, x, y)$, an equilibrium in the t-perturbed game, $(x(t), y(t))$, is entirely characterized by $\partial p_A(t, x(t), y(t))/\partial x = 0$, $\partial p_B(t, x(t), y(t))/\partial x = 0$. Performing the differentiation explicitly, we come to the next characterization of the equilibria in the t-perturbed game.

Lemma 2 *A mixed strategy pair $(x(t), y(t)) \in \text{int } S$ is an equilibrium in the t-perturbed game if and only if it solves the algebraic equation*

$$t(A^T y + g_1) + \sum_{i=1}^{n-1} \frac{e_i}{x_i} - \frac{\sum_{i=1}^{n-1} e_i}{1 - \sum_{i=1}^{n-1} x_i} + \psi_1 = 0, \tag{1}$$

$$t(Bx + h_2) + \sum_{j=1}^{m-1} \frac{f_j}{y_j} - \frac{\sum_{i=1}^{m-1} f_j}{1 - \sum_{j=1}^{m-1} y_j} + \psi_2 = 0; \tag{2}$$

here $e_i \in R^{n-1}$ and $f_j \in R^{m-1}$ are such that $e_{i,i} = 1$, $e_{i,k} = 0$ for $k \neq i$, and $f_{j,j} = 1$, $f_{j,k} = 0$ for $k \neq j$.

We call (1), (2) the *central path* equation. The formal differentiation of (1), (2) in t results in

$$\begin{pmatrix} \dot{x} \\ \dot{y} \end{pmatrix} = -H^{-1}(t, x, y) \begin{pmatrix} A^T y + g_1 \\ Bx + h_2 \end{pmatrix}. \tag{3}$$

Here

$$H(t, x, y) = \begin{pmatrix} -D(x) & tA^T \\ tB & -D(y) \end{pmatrix}, \tag{4}$$

$$D(z) = \text{Diag}\left(\frac{1}{z_1^2} + \frac{1}{\left(1 - \sum_{i=1}^k z_i\right)^2}, \ldots, \frac{1}{z_k^2} + \frac{1}{\left(1 - \sum_{i=1}^k z_i\right)^2} \right) \tag{5}$$

($z \in R^k$); $\text{Diag}(\zeta_1, \ldots, \zeta_k)$ stands for a diagonal $k \times k$ matrix with $\zeta_1, \ldots, \zeta_k$ on the diagonal. A solution $(x(\cdot), y(\cdot))$ of the differential equation (3) with the initial condition $x(t_0) = x^0$, $y(t_0) = y^0$ will always be understood as that defined on $(0, \infty)$ and satisfying $(t, x(t), y(t)) \in \mathcal{H}$ for all $t \geq t_0$. Here and in what follows $\mathcal{H}$ is the set of all $(t, x, y) \in (0, \infty) \times \text{int } S$ such that matrix $H(t, x, y)$ is invertible.

3 General Properties of Central Paths

In what follows, $Z(t \mid \psi_1, \psi_2)$ stands for the set of all equilibria in the t-perturbed game, i.e., all $(x, y) \in \text{int } S$ satisfying (1), (2). Define a *central path starting from*

$(x_0, y_0) \in \text{int } S$ *at time* t_0 to be a solution of the differential equation (3) with the initial condition $x(t_0) = x_0$, $y(t_0) = y_0$. A central path *starting in* $Z^0 \subset \text{int } S$ *at time* t_0 is understood as that starting from some $(x_0, y_0) \in Z^0$ at t_0. We call $Z^0 \subset \text{int } S$ a *set of uniqueness* for the initial time t_0 if for every $(x_0, y_0) \in Z^0$ there exists a unique central path starting from $(x_0, y_0) \in Z^0$ at t_0. For every $t_0 \geq 0$ and every $(x_0, y_0) \in \text{int } S$ we denote by $\psi_1(t_0, x_0, y_0)$ and $\psi_2(t_0, x_0, y_0)$ such $\psi_1 \in R^{n-1}$ and $\psi_2 \in R^{m-1}$, respectively, that (1) and (2) hold with $t = t_0$, $x = x_0$ and $y = y_0$.

Note that $H^{-1}(\cdot)$ is Lipschitz on every closed subset of $\mathcal{H}$. Taking this into account, we easily arrive at the following.

Lemma 3 *Let* $Z^0 \subset \text{int } S$*, and for every* $(x_0, y_0) \in Z^0$ *there exists a solution of the differential equation (3) with the initial condition* $x(t_0) = x_0$, $y(t_0) = y_0$. *Then*

(i) Z^0 *is a set of uniqueness for the initial time* t_0,
(ii) for every $t \geq t_0$ *the central path* $(x(\cdot), y(\cdot))$ *starting from* $(x_0, y_0) \in Z^0$ *at* t_0 *satisfies* $(x(t), y(t)) \in Z(t \mid \psi_1(t_0, x_0, y_0), \psi_2(t_0, x_0, y_0))$,
(iii) a function associating to each $(t, x_0, y_0) \in (t_0, \infty) \times Z^0$ *the value* $(x(t), y(t))$ *of the central path* $(x(\cdot), y(\cdot))$ *starting from* $(x_0, y_0) \in Z^0$ *at* t_0 *is continuous.*

Let N denote the set of all equilibria in the unperturbed bimatrix game. A set $Z^0 \subset \text{int } S$ will be said *to select an equilibrium* $(\bar{x}, \bar{y}) \in N$ *at the initial time* t_0 if Z^0 is a set of uniqueness for t_0, and for every central path $(x(\cdot), y(\cdot))$ starting in Z^0 at t_0 one has $\lim_{t\to\infty}(x(t), y(t)) = (\bar{x}, \bar{y})$.

Let $d(w, Y)$ and $d(X, Y)$ stand, respectively, for the distance from an element w to a set Y, and the semidistance from a set X to Y in R^k, i.e., $d(w, Y) = \inf\{|w - y| : y \in Y\}$ and $d(X, Y) = \sup\{d(x, Y) : x \in X\}$; here and in what follows $|\cdot|$ is the Euclidean norm.

Theorem 1 *Let* Ψ *be a bounded set in* $R^{n-1} \times R^{m-1}$. *Then*

$$\lim_{t\to\infty} \sup_{(\psi_1, \psi_2) \in \Psi} d(Z(t \mid \psi_1, \psi_2), N) = 0.$$

In our proof we use the next Lemma known in the theory of analytic centers for linear programming.

Lemma 4 *(see [2]). Let* $b \in R^k$, $c \in R^l$, F *be a* $k \times l$ *matrix, and* $\mu = \max\{c^T x : Fx \leq b\} > -\infty$. *Then for every* $r \geq 0$ *it holds that* $\mu - c^T x(r) \leq (k+1)r$, *where*

$$x(r) = \text{argmax}\left\{ c^T x + r \sum_{j=1}^{k} \log(b_j - (Fx)_j) : b_j > (Fx)_j,\ j = 1, \ldots, l \right\}.$$

Corollary 1 *For every* $(x_*, y_*) \in Z(t \mid \psi_1, \psi_2)$ *we have*

$$p_A(x_*, y_*) \geq \max_{x \in S_{n-1}} p_A(x, y_*) - [n + |\psi_1|(n-1)]/t, \tag{6}$$

$$p_B(x_*, y_*) \geq \max_{y \in S_{m-1}} p_A(x_*, y) - [m + |\psi_2|(m-1)]/t. \tag{7}$$

Proof Let $c^T = y_*^T A + g_1^T$ and $\bar{x}_* = \text{argmax}\{c^T x : x \in S_{n-1}\}$. By Lemma 4 $c^T \bar{x}_* \geq \max_{x \in S_{n-1}} c^T x - n/t$. Referring to the expressions for $p_A(t, x, y)$ and $p_A(x, y)$ we derive

$$|c^T \bar{x}_* - c^T x_*| \leq \max_{x \in S_{n-1}} |\psi_1||x|/t \leq |\psi_1|(n-1)/t;$$

hence, (6) is satisfied. Similarly, (7) is proved.

Proof of Theorem 1. Assume that $d(Z(t_i|\psi_{1,i}, \psi_{2,i}), N) \geq \epsilon > 0$ for some $t_i \to \infty$ and $(\psi_{1,i}, \psi_{2,i}) \in \Psi$. Take $(x_{*i}, y_{*i}) \in Z(t_i \mid \psi_{1,i}, \psi_{2,i})$ so that $d((x_{*i}, y_{*i}), N) \geq \epsilon/2$. There exist $x_i \in S_{n-1}$, $y_i \in S_{m-1}$ and $\delta > 0$ such that either $p_A(x_{*i}, y_{*i}) < p_A(x_i, y_{*i}) - \delta$, or $p_B(x_{*i}, y_{*i}) < p_A(x_{*i}, y_i) - \delta$. For large i this is not possible due to Corollary 1.

A set $D \subset S$ will be said to *contain* the central paths starting in Z^0 at t_0 if $(x(t), y(t)) \in D$ $(t \geq t_0)$ holds for each central path $(x(\cdot), y(\cdot))$ starting in Z^0 at t_0. We set $p(x, y) = (p_A(x, y), p_B(x, y))$ and $V(D) = \{p(x, y) : (x, y) \in N \cap D\}$ where $D \subset S$. We also put $V = V(S)$.

Theorem 2 *Let Z^0 be a connected set of uniqueness for the initial time t_0, $D \subset S$ be closed, contain all central paths starting in Z^0 at t_0, and $V(D)$ be finite. Then there exists $v \in V(D)$ such that $\lim_{t \to \infty} p(x(t), y(t)) = v$ for all central paths $(x(\cdot), y(\cdot))$ starting in Z^0 at t_0.*

A proof refers to several Lemmas. The next Lemma is obvious.

Lemma 5 *There exists an increasing function $\sigma(\cdot) : (0, \infty) \mapsto (0, \infty)$ such that $\lim_{\delta \to 0} \sigma(\delta) = 0$, and for every $\delta > 0$ and every $(\bar{x}, \bar{y}) \in S$ the inequalities $p_A(\bar{x}, \bar{y}) \geq \max_{x \in S_{n-1}} p_A(x, \bar{y}) - \delta$ and $p_B(\bar{x}, \bar{y}) \geq \max_{y \in S_{m-1}} p_B(\bar{x}, y) - \delta$ imply $d(p(\bar{x}, \bar{y}), V) \leq \sigma(\delta)$.*

Lemma 6 *Let Z^0 be a set of uniqueness for the initial time t_0, $D \subset S$ be closed, contain all central paths starting in Z^0 at t_0, and $V(D)$ be finite. Then for every central path $(x(\cdot), y(\cdot))$ starting in Z^0 at t_0 there exists a $v \in V(D)$ such that $\lim_{t \to \infty} p(x(t), y(t)) = v$.*

Proof Let W be the set of all limits $\lim_{i \to \infty} p(x(t_i), y(t_i))$ where $t_i \to \infty$. By Theorem 1 $W \subset V$. Due to the closedness of D, we have $W \subset D$. Therefore $W \subset V(D)$. It remains to show that W is one-element. Assume, to the contrary, that W contains two different elements, w_1 and w_2. We have $w_1 = \lim_{i \to \infty} p(x(t_i), y(t_i))$ and $w_2 = \lim_{i \to \infty} p(x(\xi_i), y(\xi_i))$ for some $t_i, \xi_i \to \infty$. With no loss in generality, assume $t_i < \xi_i < t_{i+1}$. Set

$$\nu = \min\{|v_1 - v_2| : v_1, v_2 \in V(D), \; v_1 \neq v_2\}. \tag{8}$$

Since $V(D)$ is finite, $\nu > 0$. For large i, we have $|w_1 - p(x(t_i), y(t_i))| < \nu/4$ and $|w_2 - p(x(\xi_i), y(\xi_i))| < \nu/4$. Hence $|p(x(\tau_i), y(\tau_i)) - w_1| = \nu/2$ for some $\tau_i \in (t_i, \xi_i)$.

Then, obviously, $d(p(x(\tau_i), y(\tau_i)), V(D)) \geq \nu/2$. On the other hand, for a convergent subsequence $p(x(\tau_{i_j}), y(\tau_{i_j}))$ we have $\lim_{j\to\infty} p(x(\tau_{i_j}), y(\tau_{i_j})) \in W \subset V(D)$.

Lemma 7 *Let $(x(\cdot), y(\cdot))$ be a central path starting from (x_0, y_0) at time t_0, and $\sigma(\cdot)$ be defined as in Lemma 5. Then*

(i) for every $t \geq t_0$ there is $v(t) \in V$ such that

$$|p(x(t),y(t)) - v(t)| \leq \sigma\left(\frac{n + |\psi_1(t_0, x_0, y_0)|(n-1)}{t} + \frac{m + |\psi_2(t_0, x_0, y_0)|(m-1)}{t}\right), \tag{9}$$

(ii) $\lim_{t\to\infty} v(t) = \lim_{t\to\infty} p(x(t), y(t))$.

Proof By Lemma 3 $(x(t), y(t)) \in Z(t|\ \psi_1(t_0, x_0, y_0), \psi_2(t_0, x_0, y_0))$. Referring to Corollary 1 and Lemma 5, we come to (i). Inequality (9) implies $\lim_{t\to\infty} |p(x(t), y(t)) - v(t)| = 0$. The latter together with Lemma 6 yield (ii).

Proof of Theorem 2 Suppose the statement is untrue. In view of Lemma 6, we conclude that there are two central paths, $(x_*(\cdot), y_*(\cdot))$ and $(x_*(\cdot), y_*(\cdot))$, starting at t_0 from certain $(x_{0*}, y_{0*}) \in Z^0$ and $(x_0^*, y_0^*) \in Z^0$, respectively, and converging to different points, i.e., $w_* = \lim_{t\to\infty} p(x_*(t), y_*(t)) \in V$, $w^* = \lim_{t\to\infty} p(x^*(t), y^*(t)) \in V$, and $w_* \neq w^*$. Let $\lambda \mapsto (x_0(\lambda), y_0(\lambda)) : [0,1] \mapsto Z^0$ be a continuous function such that $(x_0(0), y_0(0)) = (x_{0*}, y_{0*})$ and $(x_0(1), y_0(1)) = (x_0^*, y_0^*)$. Since Z^0 is connected, such a function exists. So far as $(x_0(\lambda), y_0(\lambda)) \in \text{int}\, S$ for all $\lambda \in [0,1]$, there exists $c > 0$ such that $|\psi_1(t_0, (x_0(\lambda), y_0(\lambda))| < c$ and $|\psi_2(t_0, (x_0(\lambda), y_0(\lambda))| < c$ for all $\lambda \in [0,1]$. Consequently, there is $t_1 \geq t_0$ such that for arbitrary $\lambda \in [0,1]$ the right-hand side in (9) with $(x_0, y_0) = (x_0(\lambda), y_0(\lambda))$ is smaller than $\nu/4$; here ν is defined by (14). By Lemma 7 for every $\lambda \in [0,1]$ and every $t \geq t_1$ there is $v(t,\lambda) \in V$ such that

$$|p(x(t,\lambda), y(t,\lambda)) - v(t,\lambda)| < \nu/4, \tag{10}$$

and

$$\lim_{\tau\to\infty} p(x(\tau,\lambda), y(\tau,\lambda)) = \lim_{\tau\to\infty} v(\tau,\lambda); \tag{11}$$

here $(x(\cdot,\lambda), y(\cdot,\lambda))$ is the central path starting from $(x_0(\lambda), y_0(\lambda))$ at t_0. Let $t \geq t_1$. Due to the continuity of the function $\lambda \mapsto (x(t,\lambda), y(t,\lambda))$ (see Lemma 3, (iii)), there exists $\epsilon > 0$ such that for every $\lambda \in [0,1]$ and all $\mu \in [0,1]$ from $B(\lambda,\epsilon)$, the open ϵ-neighborhood of λ, it holds that $|p(x(t,\mu), y(t,\mu)) - p(x(t,\lambda), y(t,\lambda)| < \nu/4$. The latter together with (10) imply $|v(t,\mu) - v(t,\lambda)| < 3\nu/4$, which is equivalent to $v(t,\mu) = v(t,\lambda)$ (see (8)). Building a finite family of neighborhoods $B(\lambda_j,\epsilon)$, $\lambda_j \in [0,1]$, $j = 1,\ldots,k$, covering $[0,1]$, we easily obtain that $v(t,\lambda) = v(t,0)$ for all $\lambda \in [0,1]$. In particular, $v(t,1) = v(t,0)$. Now take into account that $t \geq t_1$ is arbitrary and refer to (11). We arrive at $w_* = \lim_{t\to\infty} p(x(t,1), y(t,1)) = \lim_{t\to\infty} p(x(t,0), y(t,0)) = w^*$ which contradicts the assumption.

Our general equilibrium selection theorem is as follows.

Theorem 3 *Let Z^0 be a connected set of uniqueness for the initial time t_0, $D \subset S$ be closed, contain the central paths starting in Z^0 at t_0, and $p(x_*, y_*) \neq p(x^*, y^*)$ for every different (x_*, y_*), $(x^*, y^*) \in N \cap D$. Then*

(i) Z^0 selects a unique equilibrium in $N \cap D$ at time t_0, i.e., there exists $(\bar{x}, \bar{y}) \in N \cap D$ such that for all central paths $(x(\cdot), y(\cdot))$ starting in Z^0 at t_0 it holds that $\lim_{t\to\infty}(x(t), y(t)) = (\bar{x}, \bar{y})$,
(ii) if $N \cap Z^0$ is nonempty, then $N \cap Z^0 = \{(\bar{x}, \bar{y})\}$.

Proof Suppose (i) is untrue. Then by Theorem 1 there are $t_i, \xi_i \to \infty$ such that $(x(t_i), y(t_i))$ and $(x(\xi_i), y(\xi_i))$ converge to different (x_*, y_*) and (x^*, y^*) in N. Consequently, $p(x(t_i), y(t_i))$ and $p(x(\xi_i), y(\xi_i))$ have different limits, which is not possible by Theorem 2. Thus (i) is proved. Suppose (ii) is untrue, i.e, there is $(\hat{x}, \hat{y}) \in N \cap Z^0$ different from $(\bar{x}, \bar{y})$. The central path starting from $(\hat{x}, \hat{y})$ at t_0 is necessarily constant and therefore converges to $(\hat{x}, \hat{y})$ contradicting (i).

4 The Selection of the Interior Equilibrium

We call all equilibria from int $S \cap N$ *interior*. Let Z^0_ϵ be the set of all $(x, y) \in S$ such that $\epsilon \leq x^{(i)} \leq 1 - \epsilon$, $1 \leq i \leq n - 1$, and $\epsilon \leq y^{(j)} \leq 1 - \epsilon$, $1 \leq j \leq m - 1$. We see that as ϵ is sufficiently small Z^0_ϵ covers the whole int S apart from a small neighborhood of its boundary. It is clear that for every interior equilibrium $(\bar{x}, \bar{y})$ there is $\zeta > 0$ such that $(\bar{x}, \bar{y}) \in Z^0_\zeta$.

Our final result is as follows.

Theorem 4 *Let $n = m$, A and B be invertible, there exist an interior equilibrium $(\bar{x}, \bar{y})$, and $\zeta > 0$ be such that $(\bar{x}, \bar{y}) \in Z^0_\zeta$. Then for every positive $\epsilon < \zeta/2$ there exists $t_* \geq 0$ such that Z^0_ϵ selects $(\bar{x}, \bar{y})$ at every $t_0 \geq t_*$.*

A proof is based on Theorem 3, Lemma 3 and a criterion of viability.

Following [9], we call a set $F \in \mathcal{H}$ *viable* if for every $(t_0, x_0, y_0) \in F$ there exists a solution $(x(\cdot), y(\cdot))$ of the differential equation (3) with the initial condition $x(t_0) = x_0$, $y(t_0) = y_0$, and $(t, x(t), y(t)) \in F$ for all $t \in (t_0, \infty)$. Statement (i) in Lemma 3 is obviously specified as follows.

Lemma 8 *Let $F = (t_0, \infty) \times Z^0 \subset \mathcal{H}$ be viable. Then Z^0 is a set of uniqueness for the initial time t_0 and, moreover, contains all central paths starting in Z^0 at t_0.*

A standard viability criterion reads as follows. The tangent cone to $F \subset \mathcal{H}$ at point (t, x, y), further denoted $T_F(t, x, y)$, is defined to be the set of all $(\tau, \xi) \in R^1 \times (R^{n-1} \times R^{m-1})$ such that

$$\liminf_{\mu\to+0} \frac{d((t + \mu\tau, (x, y) + \mu\xi), F)}{\mu} = 0.$$

Let $G(t, x, y)$ stand for the right-hand side of equation (3).

Lemma 9 *(see [9], Theorems 1.2.1, 1.2.3). A closed set $F \in \mathcal{H}$ is viable if and only if $(1, G(t,x,y)) \in T_F(t,x,y)$ for every $(t,x,y) \in F$.*

Lemma 10 *Let $n = m$, A and B be invertible, and there exist an interior equilibrium $(\bar{x}, \bar{y})$. Then*

(i) $\bar{x} = -B^{-1}g_1$, $\bar{y} = -(A^T)^{-1}h_2$,
(ii) for each $(t,x,y) \in \mathcal{H}$

$$G(t,x,y) = -t\begin{pmatrix} -D(x)/t & A^T \\ B & -D(y)/t \end{pmatrix}^{-1}\begin{pmatrix} 0 & A^T \\ B & 0 \end{pmatrix}\begin{pmatrix} x-\bar{x} \\ y-\bar{y} \end{pmatrix}. \tag{12}$$

Conjecture (i) follows from the definition of an equilibrium, and (ii) is easily derived from (i).

Now we come back to Lemma 8 which provides a condition sufficient for Z^0 to be a set of uniqueness. In the next Lemma, using Lemmas 9 and 10, we verify this condition for Z^0_ϵ with sufficiently small ϵ.

Lemma 11 *Let $n = m$, A and B be invertible, $(\bar{x}, \bar{y})$ be an interior equilibrium, $\zeta > 0$ be such that $(\bar{x}, \bar{y}) \in Z^0_\zeta$, and $0 < \epsilon < \zeta/2$. Then there exists $t_* \geq 0$ such that for every $t_0 \geq t_*$ the set $F = (t_0, \infty) \times Z^0_\epsilon$ lies in $\mathcal{H}$ and is viable.*

Proof Referring to (5) we easily obtain that $D(x)$ and $D(y)$ are bounded on Z^0_ϵ. Then for arbitrary $\delta > 0$ and all t greater than a sufficiently large t_* we have

$$\left|\begin{pmatrix} -D(x)/t & A^T \\ B & -D(y)/t \end{pmatrix} - \begin{pmatrix} 0 & A^T \\ B & 0 \end{pmatrix}\right| < \delta \quad ((x,y) \in Z^0). \tag{13}$$

The second matrix on the left is nondegenerate, since A and B are nondegenerate by assumption. Therefore the first matrix on the left is nondegenerate too provided δ is sufficiently small. Assuming the latter, we conclude that $H(t,y,z)$ (see (4)) is nondegenerate for all $t \geq t_*$ and all $(x,y) \in Z^0$. Consequently, $F \subset \mathcal{H}$ if $t_0 \geq t_*$. It remains to show that F is viable. Without loss in generality, assume that for the matricies inverse to those in (13) we have

$$\left|\begin{pmatrix} -D(x)/t & A^T \\ B & -D(y)/t \end{pmatrix}^{-1} - \begin{pmatrix} 0 & (A^T)^{-1} \\ B^{-1} & 0 \end{pmatrix}\right| < \delta \quad ((x,y) \in Z^0).$$

whenever $t \geq t_*$. Hence, observing (12), we get

$$G(t,x,y) = -t\left[\begin{pmatrix} x-\bar{x} \\ y-\bar{y} \end{pmatrix} + \sigma(t,x,y)\right] \quad (t \geq t_*,\ (x,y) \in Z^0_\epsilon), \tag{14}$$

where, with no loss in generality,

$$|\sigma(t,x,y)| \leq \delta \quad (t \geq t_*,\ (x,y) \in Z^0_\epsilon). \tag{15}$$

Now we turn to the viability criterion of Lemma 9. Assume $t_0 \geq t_*$. Take $(t,x,y) \in F$. Let, first, $(x,y) \in \operatorname{int} Z^0_\epsilon$. Then obviously $T_F(t,x,y) = \{1\} \times R^{n-1} \times$

R^{m-1} and hence $(1, G(t,x,y)) \in T_F(t,x,y)$. Let (x,y) belong to the boundary of Z^0_ϵ. Since $\epsilon < \zeta/2$, we have $(\bar{x},\bar{y}) + \sigma \in Z^0_\epsilon$ for all $\sigma \in R^{n-1} \times R^{m-1}$ such that $|\sigma| < \zeta/2$. Assuming $\delta < \zeta/2$ and taking into account (14) and (15), we get $(x,y) + G(t,x,y)/t \in Z^0_\epsilon$. So far as Z^0_ϵ is convex and $(x,y) \in Z^0_\epsilon$, we have $(x,y) + \mu G(t,x,y)/t \in Z^0_\epsilon$ for all $\mu \in [0,1]$. Hence for all $\mu \in [0,1]$ it holds that $(t,x,y) + (\mu, \mu G(t,x,y)/t) \in (t,\infty) \times Z^0 \subset F$, or $d((t,(x,y)) + \mu(1, G(t,x,y)), F) = 0$. Therefore $(1, G(t,x,y)) \in T_F(t,x,y)$.

Proof of Theorem 4 By Lemma 11 there exists $t_* \geq 0$ such that for every $t_0 \geq t_*$ set $F = (t_0, \infty) \times Z^0_\epsilon$ lies in $\mathcal{H}$ and is viable. Take $t_0 \geq t_*$. By Lemma 8 Z^0_ϵ is a set of uniqueness for t_0. This set is convex and consequently connected. Since F is viable, Z^0_ϵ contains the central paths starting in Z^0_ϵ at t_0. Note that Z^0_ϵ is closed. Hence by Theorem 3, (ii), $N \cap Z^0_\epsilon$ contains a single equilibrium. This equilibrium is $(\bar{x},\bar{y})$, since $(\bar{x},\bar{y}) \in Z^0_\zeta \subset Z^0_\epsilon$. By Theorem 3, (i), Z^0_ϵ selects $(\bar{x},\bar{y})$ at time t_0.

References

[1] Zangwill, W.I. and Garcia, C.B. (1981) *Pathways to Solutions, Fixed Points and Equilibria*. Prentice-Hall, Inc., Englewood Cliffs, NJ 07632.

[2] Sonnevend, G. (1986) An "analytic center" for polyhedrons and new classes of global algorithms for linear (smooth convex) programming, in *Proc. 12th Conference on System Modeling and Optimization*, Budapest, 1985, Lecture Notes in Control and Inform. Sci., vol. 84. Springer, Berlin, 866–876.

[3] Sonnevend, G., Stoer, J. and Zhao, G. (1991) On the complexity of following the central path in linear programs. *Mathematical Programming*, ser. B, 527–553.

[4] Harsányi, J.C. (1982) Oddness of the numbers of equilibrium points: a new proof, in: *Papers in Game Theory* (Harsanyi, J.C., ed.), Theory and Decision Library Series, vol. 28, Dodrecht, Boston and London, 96–111.

[5] Harsányi, J.C. and Selten, R. (1988) *A General Theory of Equilibrium Selection in Games*. MIT Press, Cambridge, Mass.

[6] Brown, G.W. (1991) Iterative solutions of games by fictitious play, in *Activity Analysis of Production and Allocation* (Koopmans, T.C., ed.). Wiley, New York.

[7] Young, H.P. (1993) Evolution of conventions. *Econometrica*, **61**, 57–84.

[8] Hofbauer, J. and Sigmund, K. (1988) *The Theory of Evolution and Dynamical Systems*, London Math. Soc. Students Texts, vol. 7. Cambridge Univ. Press., Cambridge.

[9] Aubin, J.-P. (1991) *Viability Theory*. Birkhäuser, Boston.

14 ANALYTIC SOLUTIONS FOR EVOLUTIONARY NONZERO SUM GAMES*

A.M. TARASYEV

Institute for Applied Systems Analysis, Laxenburg, Austria
Institute of Mathematics and Mechanics, Ural Branch of Russian Academy of Sciences, Ekaterinburg Russia

A nonzero sum evolutionary game between two coalitions of competitors is considered. Repeated random contacts of players form a dynamical control process with infinite horizon. Payoffs to the coalitions are defined as the limits of the average incomes of their members as time goes to infinity. A dynamical Nash equilibrium is designed in a class of feedback controls. Equilibrium feedbacks are constructed analytically within a framework of the theory of viscosity (minmax) solutions of Hamilton-Jacobi equations. The resulting equilibrium dynamics is in a sense more favorable for each coalition than the classical replicator dynamics (see [3]).

Introduction

We study an evolutionary 2×2 nonzero sum game between two coalitions of competitors. The analysis is based on nonantagonistic game-differential principles [7], and a dynamical approach to game-evolutionary equilibria suggested in [10, 11]. We construct a dynamical Nash equilibrium composed of guaranteeing payoff-maximizing feedbacks and show that the trajectories under these feedbacks give, in the long run, payoffs no worse than those reached in a static Nash equilibrium point.

Dynamical game interactions are relevant to both differential (see [5, 8, 9]) and evolutionary (see [1, 3, 4, 6]) game-theoretical models. Repeated random contacts between the players can be viewed as a dynamical control process in which the state vector is composed of the proportions of players' groups adopting different

* The work has been partially supported by the International Science Foundation (NME000, NME300) and by the Russian Fund for Fundamental Researches (93-011-16032, 96-01-00219, 97-01-00161).

strategies (equivalently, the probabilities of the strategies) and control parameters regulate the flows between these groups. The process evolves over an infinite interval of time. Current incomes of the players are specified in payoff matrixes. A current income of a coalition is defined as the average of the current payoffs to its members. A long-term benefit of a coalition can be defined in different ways. In [14] game-evolutionary processes with discounted integral payoffs were analyzed. Another approach treats a long-term benefit of a coalition as the limit of its current incomes as time tends to infinity (see [10, 11]). In this paper we hold this viewpoint.

We introduce a notion of a dynamical Nash equilibrium in a class of feedback controls. Feedbacks driving the coalitions to the classical "punishment" solutions in static bimatrix games give a natural and elementary example of a dynamical Nash equilibrium. The nature of a punishment feedback is antagonistic: it is aimed at minimizing the payoff of the opposing coalition but not maximizing the own one.

We propose another solution which provides a better (at least not worse) long-term result. Our approach originates from theory of positional differential games and rests on the idea of optimal guaranteeing feedbacks in the associated zero sum games (see [7–9]). We define relevant zero sum games and study them within the framework of the theory of viscosity (minimax) solutions of Hamilton-Jacobi equations (see [2, 12]). The analysis results in analytic expressions for the value functions and optimal feedbacks. The optimal feedbacks are characterized by switching lines generated, in turn, by the structure of the value functions. We show that the equilibrium trajectories generated under the optimal feedbacks stay, in the long run, within a domain in which the current payoffs to each coalition are better (no worse) than the payoff at a static Nash equilibrium point.

1 Evolutionary Game: Dynamical Equilibria

Let a dynamics of two coalitions of competitors, each adopting one of two admissible strategies, be described by

$$\begin{aligned} \dot{x} &= -x + u, \\ \dot{y} &= -y + v. \end{aligned} \tag{1.1}$$

Parameter x, $0 \leq x \leq 1$, is treated as a probability of the fact that an individual picked randomly from coalition 1 holds strategy 1 (respectively, $(1-x)$ is the probability of the fact that the individual holds strategy 2). An equivalent interpretation for x is the proportion of the individuals holding strategy 1 in coalition 1. Parameter y, $0 \leq y \leq 1$, associated with coalition 2 has a similar sense. Control parameters u and v satisfy the restrictions $0 \leq u \leq 1$, $0 \leq v \leq 1$. They regulate interflows between the groups of individuals holding strategies 1 and 2. One can interpret u and v as signals for the individuals to change or not change their strategies. For example, $u = 0$ ($v = 0$) signals "change strategy 1 to strategy 2", $u = 1$

($v = 1$) signals "change strategy 2 to strategy 1" and $u = x$ ($v = y$) signals "do not change the strategies". The game dynamics (1.1) was justified in detail in [14].

We assume that payoffs to an individual from coalition 1 are characterized by a payoff matrix $A = \{a_{ij}\}$, and payoffs of an individual from coalition 2 by a payoff matrix $B = \{b_{ij}\}$. We define payoffs to coalitions 1 and 2 at time t, $g_A(x(t), y(t))$ and $g_B(x(t), y(t))$, as the mathematical expectations of the individual payoffs. We have

$$g_A(x(t), y(t)) = C_A x(t)y(t) - \alpha_1 x(t) - \alpha_2 y(t) + a_{22},$$
$$g_B(x(t), y(t)) = C_B x(t)y(t) - \beta_1 x(t) - \beta_2 y(t) + b_{22},$$

where (see, e.g., [15])

$$C_A = a_{11} - a_{12} - a_{21} + a_{22}, \quad \alpha_1 = a_{22} - a_{12}, \quad \alpha_2 = a_{22} - a_{21},$$
$$C_B = b_{11} - b_{12} - b_{21} + b_{22}, \quad \beta_1 = b_{22} - b_{12}, \quad \beta_2 = b_{22} - b_{21}.$$

Following [10], we define long-term benefits of the coalitions, J_A^∞, J_B^∞, as multifunctions whose values are locked between the lower and upper limits of the payoffs gained along trajectories $(x(\cdot), y(\cdot))$ of system (1.1):

$$J_A^\infty = [J_A^-, J_A^+], \; J_A^- = \liminf_{t\to\infty} g_A(x(t), y(t)), \; J_A^+ = \limsup_{t\to\infty} g_A(x(t), y(t)), \tag{1.2}$$

$$J_B^\infty = [J_B^-, J_B^+], \; J_B^- = \liminf_{t\to\infty} g_B(x(t), y(t)), \; J_B^+ = \limsup_{t\to\infty} g_B(x(t), y(t)). \tag{1.3}$$

Let us consider a differential nonzero sum game with dynamics (1.1) and multivalued payoffs (1.2), (1.3). For this game, following [7], we introduce a notion of a dynamical Nash equilibrium in the class of closed-loop strategies $U = u(t, x, y, \varepsilon)$, $V = v(t, x, y, \varepsilon)$ (here ε is a small positive accuracy parameter playing a technical role). Below, S stands for the unit square $[0, 1] \times [0, 1]$. A trajectory of system (1.1) originating from $(x_0, y_0) \in S$ under feedbacks $U = u(t, x, y, \varepsilon)$, $V = v(t, x, y, \varepsilon)$ is defined (see [8, 9]) to be a continuous function $(x(\cdot), y(\cdot)) : (0, \infty) \mapsto S$ satisfying the following condition: there are a sequence of Euler splines, $(x_j(\cdot), y_j(\cdot))$, with time steps $\delta_j \to 0$ and a sequence of accuracy parameters $\varepsilon_j \to 0$, such that $(x(\cdot), y(\cdot))$ is the uniform limit of $(x_j(\cdot), y_j(\cdot))$ on every bounded interval. An Euler spline $(x_j(\cdot), y_j(\cdot))$ is defined as a step-by-step solution of a sequence of differential equations

$$\dot{x}_j(t) = -x_j(t) + u(t_k^j, x(t_k^j), y(t_k^j), \varepsilon_j), \; \dot{y}_j(t) = -y_j(t) + v(t_k^j, x(t_k^j), y(t_k^j), \varepsilon_j),$$
$$t \in (t_k^j, t_{k+1}^j), \; t_{k+1}^j - t_k^j = \delta_j, \; k = 0, 1, \ldots, \; x_j(0) = x_0, \; y_j(0) = y_0.$$

The set of all trajectories of system (1.1) originating from $(x_0, y_0) \in S$ under feedbacks U, V will further be denoted as $X(x_0, y_0, U, V)$. It can be easily proved that $X(x_0, y_0, U, V)$ is nonempty for every $(x_0, y_0) \in S$ and every pair of feedbacks U, V.

Definition 1.1 *Let $(x_0, y_0) \in S$. A pair of feedbacks U^0, V^0 is called a dynamical Nash equilibrium for the initial position (x_0, y_0) if for any feedbacks U and V the* *following condition is satisfied: for all* $(x_1(\cdot), y_1(\cdot)) \in X(x_0, y_0, U, V^0)$, *all* $(x_2(\cdot), y_2(\cdot)) \in X(x_0, y_0, U^0, V)$, *and all* $(x^0(\cdot), y^0(\cdot)) \in X(x_0, y_0, U^0, V^0)$, *it holds that*

$$J_A^-(x^0(\cdot), y^0(\cdot)) \geq J_A^+(x_1(\cdot), y_1(\cdot)), \quad J_B^-(x^0(\cdot), y^0(\cdot)) \geq J_B^+(x_2(\cdot), y_2(\cdot)).$$

Following [7], we shall compose U^0, V^0 with the help of optimal feedbacks in zero sum type differential games Γ_A and Γ_B. In game Γ_A coalition 1 maximizes the infimum of J_A^- (see (1.2)) choosing a feedback U, and coalition 2 minimizes the supremum of J_A^+ (see (1.2)) choosing a feedback V. Symmetrically, in game Γ_B coalition 1 minimizes the infimum of J_B^+ (see (1.3)) choosing a feedback U, and coalition 2 maximizes the infimum of J_B^- (see (1.3)) choosing a feedback V. In the terminology of [8, 9], in games Γ_A and Γ_B the coalitions solve problems of guaranteed maximization and minimization. Games Γ_A and Γ_B have the antagonistic nature and are "nearly" zero sum (they would be zero sum if $J_A^- = J_A^+$ and $J_B^- = J_B^+$ were to hold).

Let us make some agreements on the notations. We shall denote u_A^0 and v_B^0 feedbacks solving, respectively, the problems of guaranteed maximization of J_A^- and J_B^- in games Γ_A and Γ_B. These feedbacks will be called positive since they maximize (with the guarantee) the long-term coalitions' benefits. By u_B^0 and v_A^0 will be denoted feedbacks mostly unfavorable for the opposite coalitions, i.e., minimizing (with the guarantee) payoffs J_B^+ and J_A^+, respectively. These feedbacks will be called punishment feedbacks.

Simple solutions of games Γ_A and Γ_B follow from classical theory of bimatrix games. In what follows, we assume for the definiteness that

$$C_A > 0, \quad 0 < x_A = \frac{\alpha_2}{C_A} < 1, \quad 0 < y_A = \frac{\alpha_1}{C_A} < 1,$$

$$C_B < 0, \quad 0 < x_B = \frac{\beta_2}{C_B} < 1, \quad 0 < y_B = \frac{\beta_1}{C_B} < 1.$$

In these assumptions, (x_B, y_A) is a single Nash equilibrium point in the bimatrix game with the payoff matrixes A and B (see, e.g., [15]). We call (x_B, y_A) the static Nash equilibrium. One can prove the following statement.

Proposition 1.1 *Differential games* Γ_A *and* Γ_B *have constant value functions*

$$v_A = \frac{a_{22}C_A - \alpha_1\alpha_2}{C_A}, \quad v_B = \frac{b_{22}C_B - \beta_1\beta_2}{C_B}, \tag{1.4}$$

respectively. Control $u_A^0(x)$ *equal* 1 *for* $x \leq x_A$ *and* 0 *for* $x > x_A$, *and control* $v_B^0(y)$ *equal* 1 *for* $y \leq y_B$ *and* 0 *for* $y > y_B$ *is a pair of positive feedbacks. Control* $u_B^0(x)$ *equal* 1 *for* $x \leq x_B$ *and* 0 *for* $x > x_B$, *and control* $v_A^0(y)$ *equal* 1 *for* $y \leq y_A$ *and* 0 *for* $y > y_A$ *is a pair of punishment feedbacks. The positive feedbacks* u_A^0, v_B^0 *lead trajectories* $(x(\cdot), y(\cdot))$ *of system* (1.1) *to point* (x_A, y_B) *ensuring* $J_A^\infty(x(\cdot), y(\cdot)) = \{v_A\}$, $J_B^\infty(x(\cdot), y(\cdot)) = \{v_B\}$. *The punishment feedbacks* u_B^0, v_A^0 *lead trajectories* $(x(\cdot), y(\cdot))$ of system (1.1) to the static Nash equilibrium (x_B, y_A) also ensuring $J_A^\infty(x(\cdot), y(\cdot)) = \{v_A\}$, $J_B^\infty(x(\cdot), y(\cdot)) = \{v_B\}$.

Dynamical Nash equilibria can be designed by pasting together positive and punishment feedbacks u_A^0 and u_B^0, and positive and punishment feedbacks v_B^0 and v_A^0. Take an initial position $(x_0, y_0) \in S$ and an accuracy parameter $\varepsilon > 0$. Choose a trajectory $(x^0(\cdot), y^0(\cdot)) \in X(x_0, y_0, u_A^0, v_B^0)$ generated by positive feedbacks u_A^0 and v_B^0. Let $T_\varepsilon > 0$ be such that $g_A(x^0(t), y^0(t)) \geq J_A^-(x^0(\cdot), y^0(\cdot)) - \varepsilon$ and $g_B(x^0(t), y^0(t)) \geq J_B^-(x^0(\cdot), y^0(\cdot)) - \varepsilon$ for $t \geq T_\varepsilon$ Let $u_A^\varepsilon(t)$ and $v_B^\varepsilon(t)$ be over-time realizations of the feedbacks u_A^0, v_B^0 in an Euler spline $(x_\varepsilon(\cdot), y_\varepsilon(\cdot))$ such that $\|(x^0(t), y^0(t)) - (x_\varepsilon(t), y_\varepsilon(t))\| < \varepsilon$ for $t \in [0, T_\varepsilon]$. Using a technique of [7] one can prove the following.

Proposition 1.2 *Feedbacks $U^0 = u^0(t, x, y, \varepsilon)$, $V^0 = v^0(t, x, y, \varepsilon)$ pasting together positive feedbacks u_A^0, v_B^0 and punishment feedbacks u_B^0, v_A^0 in accordance with*

$$U^0 = u^0(t, x, y, \varepsilon) = \begin{cases} u_A^\varepsilon(t) & \text{if } \|(x, y) - (x_\varepsilon(t), y_\varepsilon(t))\| < \varepsilon \\ u_B^0(x, y) & \text{otherwise} \end{cases}, \tag{1.5}$$

$$V^0 = v^0(t, x, y, \varepsilon) = \begin{cases} v_B^\varepsilon(t) & \text{if } \|(x, y) - (x_\varepsilon(t), y_\varepsilon(t))\| < \varepsilon \\ v_A^0(x, y) & \text{otherwise} \end{cases} \tag{1.6}$$

is a dynamical Nash equilibrium for the initial position (x^0, y^0).

Each of the positive feedbacks described in Proposition 1.1 is inflexible in the sense that it does not change controls along a trajectory until the latter reaches some critical line, and presses the system to this line afterwards. Below we construct more flexible positive feedbacks that generate trajectories $(x^{fl}(\cdot), y^{fl}(\cdot))$ attracted to positions better (at least no worse) in both criteria, J_A^∞ and J_B^∞, than the limiting points (x_A, y_B) under the positive feedbacks characterized in Proposition 1.1.

In Section 2 we solve auxiliary zero sum games with terminal payoffs. In Section 3 we construct lower envelopes of the value functions in the terminal games and obtain solutions for the multiterminal games Γ_A and Γ_B. In Section 4 we derive flexible positive feedbacks from the structure of value functions in the multiterminal games.

2 Analytic Solutions of Terminal Problems

Let us introduce a terminal zero sum differential game $\Gamma_A(T)$ with dynamics (1.1) and payoff $g_A(x(T), y(T))$ maximized and minimized by coalitions 1 and 2, respectively. Here T is a fixed terminal instant. This game has a value function w_1 (see [9]):

$$\begin{aligned} w_1(T, t_0, x_0, y_0) &= \max_U \min_{(x_1(\cdot), y_1(\cdot))} g_A(x_1(T), y_1(T)) \\ &= \min_V \max_{(x_2(\cdot), y_2(\cdot))} g_A(x_2(T), y_2(T)). \end{aligned} \tag{2.1}$$

Here (x_0, y_0) is an arbitrary initial state in S, U and V run through the sets of all feedbacks of coalitions 1 and 2, respectively; trajectory $(x_1(\cdot), y_1(\cdot))$ originates from (x_0, y_0) and corresponds to feedback U and an arbitrary open-loop control $v(t)$ of coalition 2; trajectory $(x_2(\cdot), y_2(\cdot))$ originates from (x_0, y_0) and corresponds to feedback V and an arbitrary open-loop control $u(t)$ of coalition 1. The value function w_1 satisfies the principle of dynamical programming. At points where w_1 is differentiable this principle turns into a first order partial differential equation of the Hamilton-Jacobi type:

$$\frac{\partial w_1}{\partial t} - \frac{\partial w_1}{\partial x}x - \frac{\partial w_1}{\partial y}y + \max\left\{0, \frac{\partial w_1}{\partial x}\right\} + \min\left\{0, \frac{\partial w_1}{\partial y}\right\} = 0. \tag{2.2}$$

Obviously the next terminal boundary condition is satisfied:

$$w_1(T, T, x, y) = g_A(x, y). \tag{2.3}$$

We shall find the value function w_1 (2.1) in game $\Gamma_A(T)$ as a generalized (minimax, viscosity) solution of the terminal boundary value problem (2.2), (2.3). A criterion for a (generally, nondifferentiable) w_1 to be a generalized solution (see [2, 12]) requires that the boundary condition (2.3) is satisfied and the following differential inequalities for conjugate derivatives D^*w_1 and D_*w_1 (replacing the Hamilton-Jacobi differential equality (2.2)) hold:

$$D^*w_1(T, t, x, y)|(s) \geq H(x, y, s),$$
$$D_*w_1(T, t, x, y)|(s) \leq H(x, y, s),$$
$$(t, x, y) \in (t_0, T) \times (0, 1) \times (0, 1), \quad s = (s_1, s_2) \in R^2.$$

The conjugate derivatives D^*w_1 and D_*w_1, [13], and Hamiltonian H are given by

$$D^*w_1(T, t, x, y)|(s) = \sup_{h \in R^2}(\langle s, h\rangle - \partial_- w_1(T, t, x, y)|(1, h)), \tag{2.4}$$

$$D_*w_1(T, t, x, y)|(s) = \inf_{h \in R^2}(\langle s, h\rangle - \partial_+ w_1(T, t, x, y)|(1, h)), \tag{2.5}$$

$$H(x, y, s) = -s_1x - s_2y + \max\{0, s_1\} + \min\{0, s_2\}. \tag{2.6}$$

Here symbols $\partial_- w_1(T, t, x, y)|(1, h)$, and $\partial_+ w_1(T, t, x, y)|(1, h)$ denote the lower and upper Dini directional derivatives of w_1 at a point (t, x, y) in direction $(1, h)$, $h = (h_1, h_2) \in R^2$.

The terminal boundary value problem (2.2), (2.3) has a single (generalized) piecewise smooth solution w_1 described analytically. Function w_1 is composed of the next smooth functions φ_k, $k = 1, \ldots, 5$:

$$\varphi_1(T, t, x, y) = C_A e^{2(t-T)}xy - \alpha_1 e^{(t-T)}x - \alpha_2 e^{(t-T)}y + a_{22}, \tag{2.7}$$

$$\varphi_2(T, t, x, y) = C_A e^{2(t-T)}xy - \alpha_1 e^{(t-T)}x - \left(C_A e^{2(t-T)} + (\alpha_2 - C_A)e^{(t-T)}\right)y + \alpha_1 e^{(t-T)} + a_{12}, \tag{2.8}$$

$$\begin{aligned}\varphi_3(T,t,x,y) = {} & C_A e^{2(t-T)} xy - \left(C_A e^{2(t-T)} + (\alpha_1 - C_A) e^{(t-T)}\right) x \\ & - \left(C_A e^{2(t-T)} + (\alpha_2 - C_A) e^{(t-T)}\right) y \\ & + C_A e^{2(t-T)} + (\alpha_1 + \alpha_2 - 2C_A) e^{(t-T)} + a_{11}, \end{aligned} \tag{2.9}$$

$$\begin{aligned}\varphi_4(T,t,x,y) = {} & C_A e^{2(t-T)} xy - \left(C_A e^{2(t-T)} + (\alpha_1 - C_A) e^{(t-T)}\right) x \\ & - \alpha_2 e^{(t-T)} y + \alpha_2 e^{(t-T)} + a_{21}, \end{aligned} \tag{2.10}$$

$$\varphi_5(T,t,x,y) = \frac{a_{22} C_A - \alpha_1 \alpha_2}{C_A} = \frac{a_{11} a_{22} - a_{12} a_{21}}{C_A} = \frac{D_A}{C_A} = v_A. \tag{2.11}$$

Proposition 2.1 *The value function w_1 in game $\Gamma_A(T)$ has the form*

$$w_1(T,t,x,y) = \varphi_k(T,t,x,y) \quad \textit{if} \quad (x,y) \in D_k(T,t), \quad k = 1, \dots, 5, \tag{2.12}$$

where

$$\begin{aligned} D_1(T,t) &= \{(x,y): \ x_2(T,t) \le x \le 1, \ \ 0 \le y \le y_2(T,t)\}, \\ D_2(T,t) &= \{(x,y): \ x_1(T,t) \le x \le 1, \ \ y_2(T,t) \le y \le 1\}, \\ D_3(T,t) &= \{(x,y): \ 0 \le x \le x_1(T,t), \ \ y_1(T,t) \le y \le 1\}, \\ D_4(T,t) &= \{(x,y): \ 0 \le x \le x_2(T,t), \ \ 0 \le y \le y_1(T,t)\}, \\ D_5(T,t) &= \{(x,y): \ x_1(T,t) \le x \le x_2(T,t), \ \ y_1(T,t) \le y \le y_2(T,t)\}, \end{aligned} \tag{2.13}$$

$$x_1(T,t) = \max\left\{0, 1 - \left(1 - \frac{\alpha_2}{C_A}\right) e^{(T-t)}\right\}, \quad x_2(T,t) = \min\left\{1, \frac{\alpha_2}{C_A} e^{(T-t)}\right\},$$

$$y_1(T,t) = \max\left\{0, 1 - \left(1 - \frac{\alpha_1}{C_A}\right) e^{(T-t)}\right\}, \quad y_2(T,t) = \min\left\{1, \frac{\alpha_1}{C_A} e^{(T-t)}\right\}.$$

One can prove the Proposition through a direct verification of the generalized solution criterion (2.3)–(2.5) for the Hamilton-Jacobi boundary value problem (2.2), (2.3).

3 Value Functions in Multiterminal Games

In this section we utilize the smooth φ_k components (2.7)–(2.11) of the value function w_1 (2.12) for building the value function in a multiterminal zero sum game $\Gamma_A(\infty)$. In game $\Gamma_A(\infty)$ the payoff

$$G_A(x(\cdot), y(\cdot)) = \inf_{0 \le t < \infty} g_A(x(t), y(t)) \tag{3.1}$$

is maximized by coalition 1 and minimized by coalition 2. Game $\Gamma_A(\infty)$ has a value (see [9])

$$\begin{aligned} w_A(x_0, y_0) &= \sup_U \inf_{(x_1(\cdot), y_1(\cdot))} G_A(x_1(\cdot), y_1(\cdot)) \\ &= \inf_V \sup_{(x_2(\cdot), y_2(\cdot))} G_A(x_2(\cdot), y_2(\cdot)). \end{aligned} \tag{3.2}$$

Here trajectories $(x_1(\cdot), y_1(\cdot))$ $(x_2(\cdot), y_2(\cdot))$ of system (1.1) originate from (x_0, y_0) and correspond, respectively, to feedbacks U and V of coalitions 1 and 2 and arbitrary open-loop controls of their rivals. One can verify the following properties of the value function w_A.

Proposition 3.1 *For all $(x, y) \in S$ it holds that $w_A(x, y) \leq g_A(x, y)$.*

Proposition 3.2 *The value function w_A in game $\Gamma_A(\infty)$ is uniquely determined by the differential inequalities*

$$D_* w_A(x, y)|(s) \leq H(x, y, s) \ ((x, y) \in S, \quad s \in R^2), \tag{3.3}$$

$$D^* w_A(x, y)|(s) \geq H(x, y, s) \ ((x, y) \in S, \ w_A(x, y) < g_A(x, y), \ s \in R^2), \tag{3.4}$$

and the boundary condition (2.3). At points $(x, y) \in S$, where $w_A(x, y) < g_A(x, y)$ and w_A is differentiable, inequalities (3.3) and (3.4) turn into a stationary Hamilton-Jacobi equality

$$-\frac{\partial w_A}{\partial x} x - \frac{\partial w_A}{\partial y} y + \max\left\{0, \frac{\partial w_A}{\partial x}\right\} + \min\left\{0, \frac{\partial w_A}{\partial y}\right\} = 0. \tag{3.5}$$

We shall describe the value function w_A analytically using the lower envelopes of functions φ_1 (2.7) and φ_3 (2.9) parametrized by $s = t - T$. To construct ψ_A^1, the lower envelope of φ_1, we compute the derivative of φ_1 with respect to s, let it equal zero, find a root of the obtained equation, and substitute this root into φ_1. These manipulations result in

$$\psi_A^1(x, y) = a_{22} - \frac{(\alpha_1 x + \alpha_2 y)^2}{4 C_A x y}.$$

Similarly we find ψ_A^2, ψ_A^3 and ψ_A^4, the lower envelopes of φ_2, φ_3 and φ_4, respectively:

$$\psi_A^2(x, y) = a_{11} - \frac{((C_A - \alpha_1)(1 - x) + (C_A - \alpha_2)(1 - y))^2}{4 C_A (1 - x)(1 - y)},$$

$$\psi_A^3(x, y) = C_A x y - \alpha_1 x - \alpha_2 y + a_{22},$$

$$\psi_A^4(x, y) = \frac{a_{22} C_A - \alpha_1 \alpha_2}{C_A} = v_A.$$

An analytic description of the value function w_A is as follows.

Proposition 3.3 *One has*

$$w_A(x, y) = \psi_A^i(x, y) \quad \textit{if} \quad (x, y) \in E_A^i, \quad i = 1, \ldots, 4, \tag{3.7}$$

where

$$E_A^1 = \left\{(x, y) \in [0, 1] \times [0, 1] : \quad \frac{\alpha_2}{C_A} \leq x \leq 1, \quad \frac{\alpha_1 x}{2 C_A x - \alpha_2} \leq y \leq \frac{\alpha_1}{\alpha_2} x\right\},$$

$$
\begin{aligned}
E_A^2 &= \left\{(x,y) \in [0,1] \times [0,1] : \quad 0 \le x \le \frac{\alpha_2}{C_A}, \quad h_1(x) \le y \le h_2(x)\right\},\\
E_A^3 &= E_A^{31} \cup E_A^{32},\\
E_A^{31} &= \left\{(x,y) \in [0,1] \times [0,1] : \quad \frac{\alpha_2}{C_A} \le x \le 1, \quad 0 \le y \le \frac{\alpha_1 x}{2C_A x - \alpha_2}\right\},\\
E_A^{32} &= \left\{(x,y) \in [0,1] \times [0,1] : \quad 0 \le x \le \frac{\alpha_2}{C_A}, \quad h_2(x) \le y \le 1\right\},\\
E_A^4 &= E_A^{41} \cup E_A^{42}, \qquad\qquad (3.7)\\
E_A^{41} &= \left\{(x,y) \in [0,1] \times [0,1] : \quad 0 \le x \le \frac{\alpha_2}{C_A}, \quad 0 \le y \le h_1(x)\right\},\\
E_A^{42} &= \left\{(x,y) \in [0,1] \times [0,1] : \quad \frac{\alpha_2}{C_A} \le x \le 1, \quad \frac{\alpha_1}{\alpha_2} x \le y \le 1\right\},\\
h_1(x) &= \frac{(C_A - \alpha_1)}{(C_A - \alpha_2)}(x-1) + 1, \quad h_2(x) = \frac{(C_A - \alpha_1)(x-1)}{2C_A(1-x) - (C_A - \alpha_2)} + 1.
\end{aligned}
$$

The Proposition is proved through a direct verification of the boundary condition (2.3) and the differential inequalities (3.3), (3.4) providing a criterion for w_A to be the value function in the differential game $\Gamma_A(\infty)$.

4 Equilibrium Feedbacks and Trajectories

Let u_A^{fl} be a feedback of coalition 1 that solves the problem of guaranteed maximization of payoff G_A (3.1) in the zero sum differential game $\Gamma_A(\infty)$. Later on we shall treat u_A^{fl} as a positive feedback u_A^0 in game Γ_A and thus, in agreement with Proposition 1.2, arrive at a U^0 component of a dynamical Nash equilibrium.

It is not difficult to verify that the partial derivative $\partial w_A / \partial x$ of the value function w_A changes its sign on a line K_A characterized below. Correspondingly (see, e.g., [9]), the optimal feedback u_A^{fl} has the following structure: $u_A^{fl}(x,y)$ is 0 if (x,y) lies on the right of K_A, and 1 if (x,y) lies on the left of K_A. This feedback produces a stationary sliding regime in x on K_A, i.e., $\dot{x} = 0$ on K_A. An accurate representation for u_A^{fl} is as follows:

$$
u_A^{fl}(x,y) = \begin{cases} 0 & \text{if } (x,y) \in D_A^1 \\ 1 & \text{if } (x,y) \in D_A^2 \\ x & \text{if } (x,y) \in K_A \end{cases}
$$

where

$$
\begin{aligned}
D_A^1 &= E_A^1 \cup E_A^{31} \cup E_A^{41} \setminus K_A, \quad D_A^2 = E_A^2 \cup E_A^{32} \cup E_A^{42} \setminus K_A,\\
K_A &= K_A^1 \cup K_A^2,\\
K_A^1 &= \left\{(x,y) \in [0,1] \times [0,1] : \quad y = -\frac{(C_A - \alpha_1)}{(C_A - \alpha_2)}(1-x) + 1, \quad y \le \frac{\alpha_1}{C_A}\right\},
\end{aligned}
$$

$$K_A^2 = \left\{(x,y) \in [0,1] \times [0,1]: \quad y = \frac{\alpha_1}{\alpha_2}x, \quad y \geq \frac{\alpha_1}{C_A}\right\}.$$

The optimal feedback $u_A^{fl}(x,y)$ guarantees that in the long run the current payoff to coalition 1 is no smaller than v_A. An accurate formulation is as follows. Set

$$T_A = \ln\left(\max\left\{\frac{C_A}{\alpha_2}, \frac{C_A}{C_A - \alpha_2}\right\}\right), \quad T_B = \ln\left(\max\left\{\frac{C_B}{\beta_2}, \frac{C_B}{C_B - \beta_2}\right\}\right).$$

Proposition 4.1 *For any trajectory $(x(\cdot), y(\cdot))$ under feedback u_A^{fl} there is a $t_s \in [0, T_A]$ such that for all $t \geq t_s$ state $(x(t), y(t))$ lies in domain E_A^4 (3.7) where the value function $w_A(x,y)$ in the differential game $\Gamma_A(\infty)$ is no smaller than v_A, the equilibrium value in the bimatrix game with the payoff matrixes* A and B. *Therefore, according to (3.2) $g_A(x(t), y(t)) \geq v_A$ for all $t \geq t_s$. In particular, the payoff $J_A^-(x(\cdot), y(\cdot))$ in the differential game Γ_A is no smaller than the value in Γ_A (see Proposition 1.1):*

$$J_A^-(x(\cdot), y(\cdot)) = \liminf_{t \to \infty} g_A(x(t), y(t)) \geq v_A.$$

The last statement in Proposition 4.1 yields that u_A^{fl} is a positive feedback of coalition 1, $u_A^{fl} = u_A^0$. Similarly, we construct a positive feedback $v_B^{fl} = v_A^0$ of coalition 2. The properties of the feedbacks u_A^{fl} and v_B^{fl} yield the following.

Proposition 4.2 *For every trajectory $(x^{fl}(\cdot), y^{fl}(\cdot))$ of system (1.1) generated by the feedbacks u_A^{fl}* and *v_B^{fl} it holds that $(x^{fl}(t), y^{fl}(t)) \in E^0 = E_A^4 \cap E_B^4$ and hence $g_A(x^{fl}(t), y^{fl}(t)) \geq v_A$ and $g_B(x^{fl}(t), y^{fl}(t)) \geq v_B$ for all $t \geq T^0 = \max\{T_A, T_B\}$.*

As shown in Proposition 1.2 the positive feedbacks $u_A^0 = u_A^{fl}$ and $v_B^0 = v_B^{fl}$ supplemented with some punishment feedbacks u_B^0 and v_B^0, respectively, form a dynamical Nash equilibrium (see (1.5), and (1.6)).

The behavior of the dynamically equilibrium trajectories $(x^{fl}(\cdot), y^{fl}(\cdot))$ generated by the positive feedbacks u_A^{fl} and v_B^{fl} is of special interest. A scope of theoretically admissible situations is as follows:

(i) $(x^{fl}(\cdot), y^{fl}(\cdot))$ converges to a point on the intersection of the switching lines K_A and K_B;
(ii) $(x^{fl}(\cdot), y^{fl}(\cdot))$ tends to the boundary of the square S;
(iii) $(x^{fl}(\cdot), y^{fl}(\cdot))$ tends the interior Nash equilibrium (x_B, y_A);
(iv) $(x^{fl}(\cdot), y^{fl}(\cdot))$ cycles in domain E^0;
(v) $(x^{fl}(\cdot), y^{fl}(\cdot))$ possesses a chaotic behavior in E^0.

A complete classification of the admissible behaviors of the trajectories $(x^{fl}(\cdot), y^{fl}(\cdot))$ will presumably be a theme for the further analysis.

It is interesting to compare the designed equilibrium dynamics with the classical replicator game-evolutionary dynamics (see [3]). The trajectories governed by the replicator dynamics cycle around the static Nash equilibrium (x_B, y_A). Hence the lower bounds for J_A^- and J_B^- for the long-term coalitions' benefits gained along these trajectories are smaller than, respectively, v_A and v_B, the static payoffs at (x_B, y_A). The equilibrium trajectories generated by the positive feedbacks u^{fl} and v^{fl} have J_A^- and J_B^- no smaller than v_A and v_B, respectively. In this sense the described equilibrium dynamics is more favorable for each coalition than the replicator dynamics.

References

[1] Basar, T. and Olsder, G.J. (1982) *Dynamic Noncooperative Game Theory*. Acad. Press, London, NY.

[2] Crandall, M.G. and Lions, P.-L. (1983) Viscosity solutions of Hamilton-Jacobi equations. *Trans. Amer. Math. Soc.*, **277**(4), 1–42.

[3] Hofbauer, J. and Sigmund, K. (1988) *The Theory of Evolution and Dynamic Systems*, Cambridge Univ. Press, Cambridge.

[4] Intriligator, M. (1971) *Mathematical Optimization and Economic Theory*. Prentice-Hall, N.Y.

[5] Isaacs, R. (1965) *Differential Games*. John Wiley & Sons, New York.

[6] Kaniovski, Y.M. and Young, H.P. (1994) Learning dynamics in games with stochastic perturbations. IIASA, Laxenburg, WP-94-30.

[7] Kleimenov, A.F. (1993) *Nonantagonistic Positional Differential Games*. Nauka, Ekaterinburg (in Russian).

[8] Krasovskii, N.N. (1985) *Control of a Dynamical System*. Nauka, Moscow (in Russian).

[9] Krasovskii, N.N. and Subbotin, A.I. (1988) *Game-Theoretical Control Problems*. Springer NY, Berlin.

[10] Kryazhimskii, A.V. (1994) Behavioral equilibria for a 2×2 "seller-buyer" game-evolutionary model. IIASA, Laxenburg, WP-94-131.

[11] Kryazhimskii, A.V. and Osipov Yu.S. (1995) On differential-evolutionary games. *Proceedings of the Steklov Institute of Mathematics*, **211**, 234–261.

[12] Subbotin, A.I. (1995) Generalized solutions for first-order PDE. The dynamical optimization perspective. Birkhauser, Boston. *Systems and Control: Foundations and Applications*.

[13] Subbotin, A.I. and Tarasyev, A.M. (1985) Conjugate derivatives of the value function of a differential game *Soviet Math. Dokl.*, **32**(2), 162–166.

[14] Tarasyev, A.M. (1994) A differential model for a 2×2 evolutionary game dynamics. IIASA, Laxenburg, WP-94-63.

[15] Vorobyev, N.N. (1985) *Games Theory*. Nauka, Moscow (in Russian).

APPLICATIONS TO SOCIAL AND ENVIRONMENTAL PROBLEMS

15 OPTIMAL CONTROL OF LAW ENFORCEMENT*

GUSTAV FEICHTINGER

Vienna University of Technology, Institute for Econometrics, Operations Research and Systems Theory, Vienna, Austria

This paper reviews some recent results on the economics of crime and punishment. In particular, a dynamic extension of Becker's static setup is proposed. Moreover, the enforcement of two types of consensual crimes is illustrated, namely corruption and drug control. Social interaction (and density-dependence) imply multiple equilibria, a fact which may help to explain the variety of different levels of corruption in societies under similar conditions. Finally, it is shown how optimal enforcement policies in drug control depends on the structure of the illicit market.

1 Intertemporal Modeling in the Economics of Crime

In his pathbreaking paper *Crime and punishment: An economic approach*, Gary S. Becker [4], Nobel prize winner in economics in 1992, asked how much resources should be spent in order to prevent offenses and to convict offenders. In particular, he tries to find those public and private expenditures and punishment that minimize the total social loss from offenses. To determine the optimal amount of enforcement one has to know the damage caused by the offenses and the response of offenders to changes in law enforcement, the cost of apprehending and convicting criminals, and the impact of the nature and the amount of punishments meted out.

Becker [4] claims that his economic approach may dispense with special theories to explain criminal behavior like inheritance or psycho-social inadequacies. As one of his highlights he is able to determine the level of offenses a society should permit, i.e. how many offenders should go unpunished.

---

* The results sketched in this paper are *preliminary* ones and may be seen as part of an ongoing research project on dynamic optimization models of optimal law enforcement. Part of this paper has been presented at the joint DGOR-GMOEOR-OEGOR meeting of the German and Austrian Operations Research Societies in Passau, Germany (Symposium on Operations Research, September 1995).

Becker's approach is basically *static*. In particular, his supply function depends on the probability of conviction and the fine per offense. It is somewhat surprising that almost all of the many follow-up extensions of Becker's paper are not dynamic[1].

In reality, however, the number of offenses at a certain period depends, among other things, crucially on the enforcement in the past. Crime and its enforcement are intertemporal activities. In particular we assume that the number of offenders may be determined dynamically as follows.

To simplify the following stylized analysis we firstly identify the number of offenders with those of the offenses. (This can be done by assuming that each offender commits crimes with the same constant intensity.) Furthermore, we assume that the income y, measured in monetary terms, from an offense is constant. Denoting by u the individual utility, by p the probability that an offense is cleared by conviction per unit of time, and by f the fine, the utility expected from committing an offense is given by[2]

$$\mathbb{E}(u) = pu(y-f) + (1-p)u(y), \tag{1}$$

where $\mathbb{E}$ denotes the mathematical expectation.

Moreover, let w be the average wage rate of the competing legal activity assumed to be constant for all offenders and over time.

Then the dynamics for the offenders is assumed to be

$$\dot{N}(t) = \mathbb{E}(u) - w - apN(t) = b - p[u(y) - u(y-f)] - apN(t) \tag{2}$$

where $b = u(y) - w$ is constant which is reasonably assumed to be positive. The idea behind the differential equation (2) is that potential offenders become criminals as soon as their expected utility from a criminal activity exceeds the average legal wage rate. Note the similarity of this mechanism to the well-known replicator dynamics.

The parameter $a \geq 0$ can be interpreted either as the rate of imprisonment or as the rate of deterrence of convicted individuals. Here the term $apN(t)$ refers to the deterrence effect of apprehended offenders. It says that a proportion of the convicted offenders is deterred from commiting further crimes.

In particular, for risk neutral offenders, i.e. for a linear utility function, we have[3]

$$\dot{N} = b - pf - apN. \tag{3}$$

Note that $N(t)$ is the state variable, a and b are nonnegative parameters with $a \in [0,1]$, and p and f are functions which are either given or must be determined optimally.

---

[1] There are a few exceptions, however, e.g. Leung [11, 12]. As they consider the criminal history of individuals, but we are interested in macrodynamic extensions, these papers are of minor relevance for the following.

[2] Here it is assumed as in Becker [4] that the offender commits the crime and enjoys the benefits before he faces the possibility of punishment. Leung [11] made the more realistic assumption that at least a part of the returns is lost if the offender is caught in the act. However, since the following model has the same structure with or without this assumption we keep Becker's formulation.

[3] Here and in the rest of the paper the time argument is mostly omitted.

Moreover, the inequality

$$N(t) \geq 0 \quad \text{for all} \quad t \geq 0 \tag{4}$$

has to be satisfied.

The instruments of the law enforcement authority are the enforcement rate E and the fine f. Focusing on the first we consider the enforcement rate E as control variable. For simplicity, we set $f = 1$.

Then the social planner is faced with the problem of minimizing the discounted stream of the social loss over a given planning period. For simplicity, we choose an infinite planning horizon[4]:

$$\min_E \left\{ L = \int_0^\infty \exp(-rt)[D(N) + C(E)]dt \right\}, \tag{5}$$

where $r > 0$ denotes the time preference rate of the enforcement authority, $D(N)$ is the damage function with

$$D(0) = 0, \quad D'(N) > 0, \quad D''(N) \geq 0,$$

and $C(E)$ are the expenditures corresponding to the enforcement rate E. It is reasonable to assume

$$C(0) = 0, \quad C'(E) > 0, \quad C''(E) \geq 0.$$

Note that the convexity both of $D(N)$ and $C(E)$ makes economic sense.

The probability that an offense is cleared by conviction depends on the per capita expenditures for enforcement, $e = E/N$, with non-increasing marginal efficiency:

$$p(0) = 0, \quad p'(e) < 0, \quad p''(e) \leq 0.$$

Clearly, it must hold that $\lim_{e\to\infty} p(e) \leq 1$.

Remark For $N = 0, e = E/N$ is not defined. There are at least two ways out of this dilemma: First, redefine e as $E(N + \varepsilon)^{-1}$ with a small $\varepsilon > 0$; second, impose a state constraint $N \geq \varepsilon > 0$. In the present case, however, it can be shown that the origin in the (N, E) phase portrait behaves like a saddle point (see below and Feichtinger et al. [9]).

Together this defines a deterministic optimal control problem. The authority has to choose the enforcement rate over time such that the discounted total social loss in (5) is minimized subject to the state equation

$$\dot{N} = b - (1 + aN)p(E/N), \tag{6}$$

[4] In Becker's static approach there is a third term in the objective function, namely the social costs of punishments. However, if punishment is restricted to fines paid by offenders, this expression can be neglected as explained in Becker [4]. Note that in this case the parameter a can be interpreted as rate of deterrence of convicted persons.

the pure state constraint (4) and a given initial state $N(0) = N_o \geq 0$. The number of offenders, $N(t)$, is the state variable, and the enforcement intensity, $E(t)$, acts as the control[5].

To get an idea of the nature of the optimal solutions, we specify the functions as follows:

$$D(N) = N^2, \quad C(E) = \frac{c}{2}E^2, \quad p(e) = \begin{cases} pe & \text{for } e < 1/p \\ 1 & \text{for } e \geq 1/p\,. \end{cases}$$

The results[6] are summarized in the phase diagram exhibited in Fig. 1. The shape of the isoclines is such that there exist at most two equilibria in the (N, E) phase space.

For $a = 0$ it turns out that there exists (at most) one equilibrium which is unstable. The situation depicted in Fig. 1 illustrates the case of two equilibria $\hat{P}_1$ and $\hat{P}_2$. The

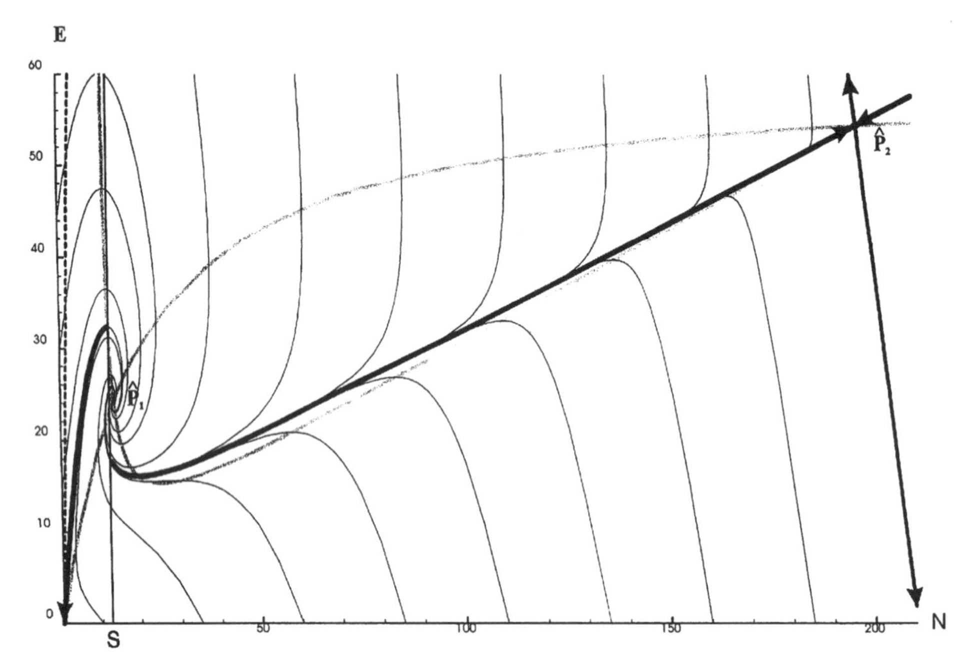

Figure 1 Phase portrait of the N-E-space ($c = 2$, $p = 0.05$, $r = 0.01$, $a = 0.05$ and $b = 0.15$). The vertical line around the 'low' equilibrium $\hat{P}_1 = (\hat{N}_1, \hat{E}_1)$ denotes the Skiba point S. The thin and grey lines are the flux and the isoclines, respectively. The thick black lines represent the optimal trajectories resulting from the existence of S, and the middle thick black lines around the 'high' equilibrium $\hat{P}_2 = (\hat{N}_2, \hat{E}_2)$ are its unstable manifolds. The dashed line indicates the border of the phase space to the left.

---

[5] Moreover, one might include the amount of the fine as additional control variable.

[6] For this we refer to Feichtinger *et al.* [9]. The applied techniques, i.e. the optimality conditions of the maximum principle and its handling, are described, e.g., in Feichtinger and Hartl [8].

left one is an unstable focus, while the right one is a saddle point. The sufficient optimality condition (concavity of the maximized Hamiltonian with respect to N) is not satisfied, and the stable branches of the saddle point path are the only candidates for the optimal solution. The suggested shape of the 'optimal' enforcement policy makes economic sense. The 'upper equilibrium' is approached from south-west and north-east, respectively. This means that for a sufficiently large initial number of offenders the corresponding 'optimal' law enforcement rate is relatively high (compared with the equilibrium enforcement rate $\hat{E}_2$, but gradually decreases. On the other hand, for a medium sized population of offenders, the crime control should be initially moderate but gradually increases to its (upper) equilibrium value $\hat{E}_2$). Note that in both cases there is a long-run steady state $\hat{N}_2$) which is asymptotically approached.

It should be noted that the conviction probability $p(e)$ is smaller than one in both equilibria.

Furthermore, it is possible to prove the existence of a Skiba point S. According to Skiba [15] there exists an interval of N-values that contains the unstable focus on which two candidate trajectories occur. One trajectory goes to the right and the other leads to the left. For $N(0) > S$ the saddle point branch winding out of $\hat{P}_1$) and converging to the upper equilibrium $\hat{P}_2$) is 'optimal'. On the other hand, for $N(0) < S$ it can be shown that the path spiraling out from the 'lower equilibrium' and leading to the origin is the 'optimal one'.

This behavior could be interpreted economically as follows: Starting with a relatively low initial stock of offenders, $N(0) < S$, it would be optimal to eradicate the offenses completely. The corresponding enforcement rate is relatively high and increases first, but decreases monotonically thereafter.

For $N(0) > S$, i.e. if the offenses exceed a certain threshold, a long run positive stock of offenders is optimal. The optimal enforcement rate increases (decreases) monotonically for $S < N(0) < \hat{N}_2(\hat{N}_2 < N(0))$. This result characterizes the optimal intertemporal trade-off between the damage generated by offenders and the expenditures to combat them.

2 Corruption and its Control

Corruption in some form or other occurs so frequently that it is hard to describe it as deviate behavior. Although the phenomenon has many dimensions including historical, political and social ones, we restrict our considerations here to the economic aspects of corruption.

What we have in mind are economically motivated illegal transactions between a member of an organization, e.g. a bureaucrat, and an agent outside of it. The bureaucrat provides an economic value, e.g. a local information or an unwarranted license, for the non-member in order to gain some economic rent.

People are aware that some degree of corruption is unavoidable, a necessary evil or — within certain limits — even useful. Public attitude requires a certain level of reported corruption before opposition is triggered. It is this threshold property as well as other nonlinearities which make corruption a phenomenon whose investigation is not only economically interesting but also worthwhile from a mathematical point of view. For an introduction into the field see Rose-Ackerman [13, 14] and the survey of Andvig [2].

The abundant literature on case studies of reported corruption in various regions of the world suggests several general questions. Mathematical models of corruption try to get answers to these open problems. Let us briefly summarize some of these questions:

- *What is the impact of corruption on welfare?* While high corruption levels reduce economic efficiency (delays, no decentralization, inability to decide), 'light bribing' might have positive economic effects (e.g. increase of service speed).
- *Why are not more people corrupt?* If corruption is economically advantageous and not too risky, why does the social norm 'to stay clean' still persist over time and space?
- Why do we observe *different levels of corruption* in comparable situations?

To answer the second question Andvig [2] proposed to include reputation into the utility of corrupt agents as Akerlof [1] did in his explanation of involuntary unemployment. According to that *'people want to be rich and famous'* the latter not being redundant; see Dawid and Feichtinger [7] for a dynamic setup of this problem.

The third problem has been attacked by taking into consideration *social interaction.* By this we mean that the utility an individual receives from a given action depends on the choices of the others in that individual's reference group. It turns out that the same kind of agents within the same kind of economic and political system may through their interaction generate several levels of corruption (compare Andvig, [2]).

3 Drug Controls

During the past several years, law enforcement in illicit drug markets has developed to a growing field in economics and planning. Caulkins [5] discusses the efficiency of crackdowns on drug dealers. Baveja *et al.* [3] analyze enforcement programs of finite duration that minimize the total costs of crackdown, subject to the constraint that the market is eliminated at the end of the program. Their main result is that the simple strategy of using the maximum available enforcement level until the market has collapsed is optimal under most circumstances.

Kort *et al.* [10] extended this analysis in various directions. First, besides the costs of enforcement the current disutility caused by dealers is included in the objective functional. Second, there is no exogenously given upper bounding of the enforcement rate. The system dynamics originates from Caulkins [5]. He specifies the rate of change of dealers with respect to time in a given market as follows:

$$\dot{N}(t) = \sigma\left[\frac{\pi Q(N(t))}{N(t)} - \left(\frac{E(t)}{N(t)}\right)^{y} - w\right], \tag{7}$$

where $N(t)$ denotes the number of dealers in the market at time $t, Q(N) = \alpha N^{\beta}$ are the total sales per period (α, β are constant demand parameters, $\beta \in [0, 1], \alpha > 0$), $E(t)$ is the enforcement (crackdown) effort at time t, σ is the adjustment speed, π denotes the profit per transaction, w is the dealer's reservation wage, and $\gamma \in (0, 1)$ denotes a parameter associated with per dealer cost of enforcement effort.

Omitting the detailed discussion of the ingredients of equation (7) we notice only that the rate of change of dealers is proportional to the difference between the utility of one dealer and his wage rate; compare the system dynamics of offenders in equation (2). While Caulkins [5] and Baveja *et al.* [3] consider risk-averse dealers, i.e. $\gamma > 1$, Kort *et al.* assume $\gamma < 1$, i.e. risk-seeking offenders.

The objective of the drug police is to minimize the discounted flow of social costs arising from the enforcement expenditures and from the disutility caused by the dealers. For mathematical convenience it is assumed that these costs are separable and linear so that the objective becomes

$$\min_{E}\left\{\int_{0}^{\infty} \exp(-rt)[E(t) + \rho N(t)]\right\} dt, \tag{8}$$

where the positive constant ρ measures the relative cost of the drugs market, subject to the Caulkins dynamics (7) for a given initial level of dealers $N(0) = N_o \geq 0$.

It should be noted that

$$N(t) \geq 0 \quad \text{for all} \quad t \in (0, \infty) \tag{9}$$

is not automatically satisfied. Its validity has to be checked as a pure state constant.

Using the maximum principle, interesting insights into the qualitative behavior of optimal enforcement paths can be obtained. In the following brief discussion we restrict ourselves to the case of a pure seller's model market and a pure buyer's market. For a more complete analysis we refer to the original paper by Kort *et al.* [10].

An illicit drug market is a pure *seller's market* if it holds that each dealer creates its own demand. Here demand is abundant in the sense that if an additional dealer enters the market, market sales will expand enough such that none of the existing dealers lose sales. In our model this means $\beta = 1$.

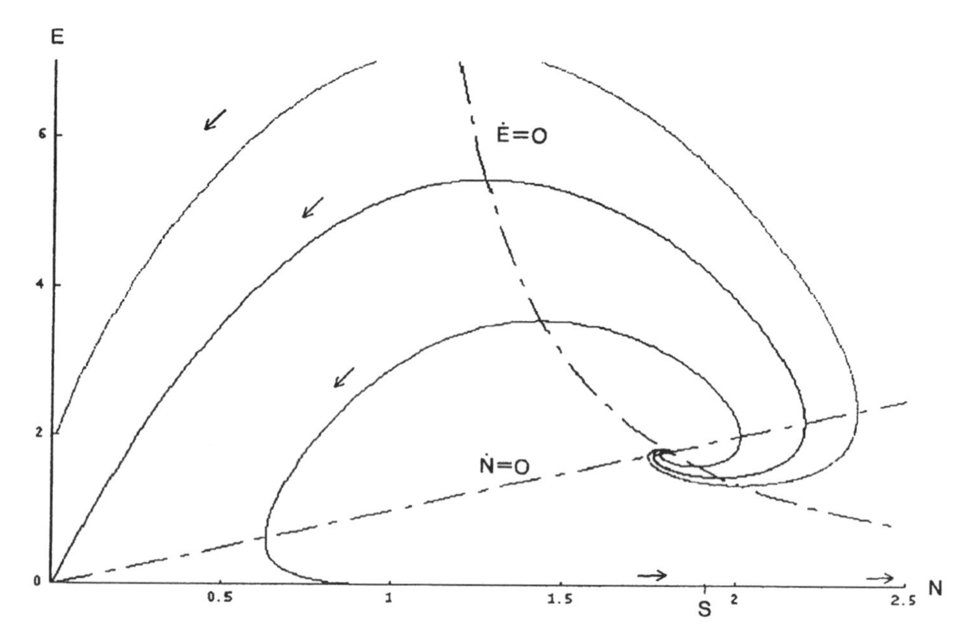

Figure 2 Candidates for an optimal path in case of a pure sellers' market ($\beta = 1$); $\gamma = 0.6$, $\sigma = 0.05$, $r = 0.05$, $\rho = 2$, $w = 10$, $\pi\alpha = 11$.

Fig. 2 presents the phase portrait of the (N, E) plane. It exhibits again a Skiba-point S such that for $N(0) > S$ a trajectory to the right is better, while for $N(0) < S$ a path leading to a market collapse is optimal (for details see Kort *et al.*, [10]).

To interpret this form of the solution it is convenient to write down the state equation (7) for $\beta = 1$ as

$$\dot{N} = \sigma\left[\pi\alpha - \left(\frac{E}{N}\right)^{y} - w\right]. \tag{10}$$

Since the burden of enforcement is shared equally among the dealers, this burden is relatively low for the individual dealer if N is large. Then, according to (10), a large enforcement rate does not prevent new dealers from entry into the market since each of them obtains a profit of $\pi\alpha$ from selling his drugs. Thus, the effectiveness of enforcement is low, and $E = 0$ for large values of N makes economic sense.

On the other hand, for $N(0) < S$ enforcement is effective enough so that it is optimal to invest resources in it. In fact, enforcement remains positive until the market collapses, i.e. $N = 0$.

A *buyer's market* is one in which sellers have no bargaining power. In such a case there is a fixed number of total sales which is to be distributed equally among the N dealers in the market. Now $\beta = 0$. The entrance of an additional dealer reduce the per capita profit.

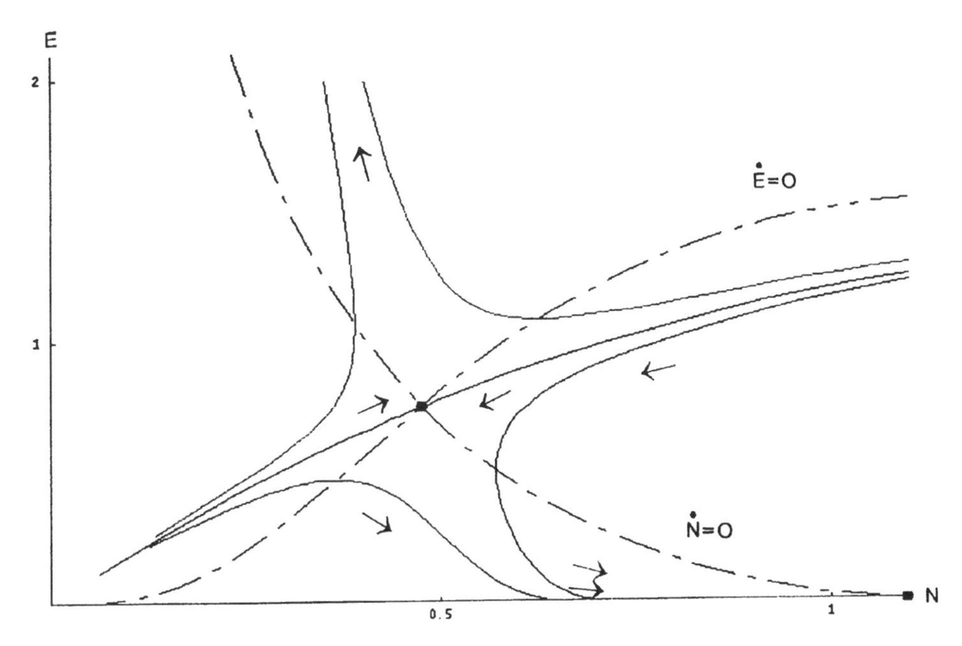

Figure 3 Candidates for an optimal path in case of a pure buyers' market ($\beta = 0$); $\gamma = 0.6$, $\sigma = 0.05$, $r = 0.05$, $\rho = 4$, $w = 1$, $\pi\alpha = 1.1$

To describe and interpret the solution we again specialize (7), but now for $\beta = 0$ as

$$\dot{N} = \sigma\left[\frac{\pi\alpha}{N} - \left(\frac{E}{N}\right)^{\gamma} - w\right]. \tag{11}$$

Contrary to the former case, here each dealer's revenue is low for large N. Since $\gamma < 1$, the revenue per dealer decreases more by large N than the burden of the crackdown felt by the individual dealer. Thus, in a buyer's market enforcement is relatively successful in fighting dealers, since a large number of dealers is unappealing to dealers. Moreover, for smaller N a market collapse is not optimal. This is because the revenue per dealer increases more than the disutility from the enforcement pressure with decreasing number of dealers. Thus, a market collapse would require an enormous amount of enforcement effort. Hence, due to the assumption that dealers are risk-seeking, the branches of the saddle-point path shown in Fig. 3 are optimal.

Summing up, we are able to say that the optimal enforcement policy depends crucially on the market structure and that the mathematically obtained crackdowns make economic sense.

For a differential game analysis of a drug control problem we refer to Dawid and Feichtinger [6].

Remark

This paper reflects the 'state of the art' some three years ago. More recent work on corruption and drug dynamics is now available. Please contact the author on that issue (e-mail address: or@e119ws1.tuwien.ac.at).

Acknowledgement

The help of the following persons is gratefully acknowledged: H. Dawid, J.L. Haunschmied, P. Kort, A. Milik and G. Tragler.

References

[1] Akerlof, G.A. (1980) A theory of social custom, of which unemployment may be one consequence. *Quarterly Journal of Economics*, **94**, 749–775.
[2] Andvig, J.C. (1991) The economics of corruption: a survey. *Studi economici*, **43**, 57–94.
[3] Baveja, A., Batta, R., Caulkins, J.P. and Karwan, M.H. (1992) Collapsing street market for illicit drugs: the benefits of being decisive. Working Paper.
[4] Becker, G.S. (1968) Crime and punishment: an economic approach. *Journal of Political Economy*, **76**, 169–217.
[5] Caulkins, J.P. (1993) Local drug markets' response to focused police enforcement. *Operations Research*, **41**, 848–863.
[6] Dawid, H. and Feichtinger, G. (1996) Optimal allocation of drug control efforts: a differential game analysis. *Journal of Optimization Theory and Applications*, **91**, 279–297.
[7] Dawid, H. and Feichtinger, G. (1996) On the persistence of corruption. *Journal of Economics*, **64**, 177–193.
[8] Feichtinger, G. and Hartl, R.F. (1986) *Optimale Kontrolle ökonomischer Prozesse: Anwendungen des Maximumprinzips in den Wirtschaftswissenschaften*. de Gruyter, Berlin.
[9] Feichtinger, G., Milik, A. and Tragler, G. (1995) Optimal dynamic law enforcement. Forschungsbericht 197 des Instituts für Ökonometrie, OR und Systemtheorie, TU Wien.
[10] Kort, P.M., Feichtinger, G., Hartl, R.F. and Haunschmied, J.L. (1996) Optimal enforcement policies (crackdowns) on a drug market. Discussion paper 9629, CentER for Economic Research, Tilburg University, Netherlands.
[11] Leung, S.F. (1991) How to make the fine fit the corporate crime? An analysis of static and dynamic optimal punishment theories. *Journal of Public Economics*, **45**, 243–256.
[12] Leung, S.F. (1995) Dynamic deterrence theory. *Economica*, **62**, 65–87.
[13] Rose-Ackerman, S. (1975) The economics of corruption. *Journal of Public Economics*, **4**, 187–203.
[14] Rose-Ackerman, S. (1978) *Corruption, a study in political economy*. Academic Press, New York.
[15] Skiba, A.K. (1978) Optimal growth with a convex–concave production function. *Econometrica*, **46**, 527–539.

16 THE CLIMATE CHANGE PROBLEM AND DYNAMICAL SYSTEMS: VIRTUAL BIOSPHERES CONCEPT

YURI M. SVIREZHEV and WERNER VON BLOH

Potsdam Institute for Climate Impact Research, D-14412 Potsdam, Germany

A dynamical model of the climate-biosphere system involving a carbon cycle is designed. A qualitative analysis is given for a model zero-dimensional in space. Three stable states of the biosphere are found: a "cold desert" with no vegetation, a "cold planet" rich in vegetation, and a "hot planet" with poor vegetation. Their stability properties can change dramatically due to an increase in the total amount of carbon caused, e.g., by anthropogenic emissions.

1 Introduction

The Climate Change Problem is very fashionable today. A huge amount of publications mask (sometimes very successfully) some simple facts that we can hardly conceive.

- How does the "Biosphera Machina" operate?
- Is the Earth Biosphere unique, or do there exist other virtual biospheres?

To answer these questions we consider the "biosphere + climate" system as a nonlinear dynamical system with multiple equilibria. Here we shall stay within a framework of a simplest modeling approach. A qualitative analysis will be given for a model zero-dimensional in space.

It is obvious that vegetation dynamics depends on the temperature, precipitation and concentration of carbon in the atmosphere. On the other hand, the temperature dynamics depends on the concentrations of carbon and water vapour in the atmosphere, and the albedo of the planetary surface. For instance, the albedo of a "white sands" desert is equal to 0.4; for a coniferous forest it is about 0.1. Our model includes the following simple submodels: global carbon cycle, global hydrological cycle, vegetation, and an equation for annual global temperature.

A few words about the history of this problem. We think that initially it was formulated (at a qualitative level, of course) by Vernadsky (1927) in the form of an idea about the interdependence between vegetation and climate (see [4]). Then Kostitzin (1935) realized this idea in the first mathematical model for the coevolution of atmosphere (climate) and biota (see [1]). It is interesting that he has obtained the "epoques glaciaires" as self-oscillations in this model. Recently Watson and Lovelock (1983) further developed Vernadsky's idea (see [5]). They considered a causal loop between the surface temperature and two types of vegetation (differing in albedo). A competition between them on a temperature ecological niche generates different spatial vegetation and temperature patterns ("Daisy World"). In 1993 Schellnhuber and von Bloh realized this model, using cellular automata, in the form of a 2-dimension structure (see [2]). This approach showed an important role of fluctuations in forming spatial patterns. In 1994 Svirezhev formulated a concept of "virtual biospheres" ([3]). According to this concept the contemporary Earth Biosphere is one of many possible (virtual) biospheres corresponding to different equilibria of a nonlinear dynamical "climate + biosphere" system. In the course of the planetary history and own evolution, this system passed through several bifurcation points in which some random factors (small perturbations) determined solution branches actually followed by the system. A driving force in this evolution could be the evolution of the "Earth green cover", which had, in turn, several bifurcation points, for instance, the appearance of terrestrial vegetation, or the change from coniferous forest to deciduous forest.

Note that this concept contradicts Vernadsky's "ergodicity axiom", according to which the contemporary Earth Biosphere is unique, irrespective of the initial and previous states.

What kind of bifurcations are admissible in this system? We can observe a change of planet's "status" from a "cold desert" to either a "cold green planet", or a "hot green planet" (the first bifurcation). Then either a "wet hot" planet covered with a tropical rain forest, or a "dry hot" planet (savannah) develops from a "green hot" planet as a result of the second bifurcation. Analogically, either "a wet cold" (a temperate forest) or "a dry cold" (steppe) planet arises from a "green cold" planet (see Fig. 1).

2 Model Description and Basic Equations

The climate of our hypothetical planet is described by a single variable, namely, the annual average temperature $T(x, t)$ on its surface. Here x is a vector of spatial coordinates. The isotropic atmosphere of the planet contains two "greenhouse" gases: carbon dioxide, CO_2, with a total amount of carbon in an atmosphere column

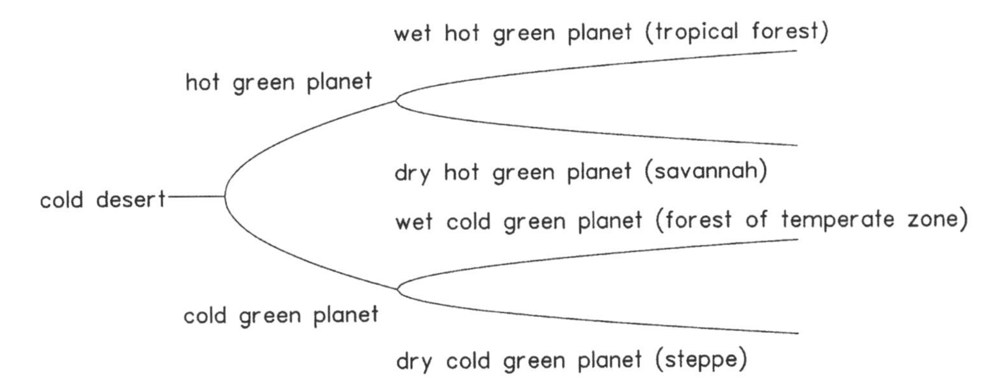

Figure 1 Bifurcation diagram for the "climate + biosphere" system.

over a surface unit being equal to $C(x, t)$, and water vapour with a total amount $W_a(x, t)$. These values can be considered as corresponding concentrations of these substances (in appropriate units). In what follows, all constant coefficients and all variables (apart from time and space coordinates) are assumed to be positive.

An equation for the temperature is

$$k\frac{\partial T}{\partial t} = D_T \nabla^2 T + S(1 - \alpha) - \sigma_{eff} T^4 - k_W W_a n. \tag{1}$$

Here k is a surface heat capacity, D_T is a thermal convectivity coefficient, S is a solar radiation coefficient, α is a surface albedo, and

$$\begin{aligned} \sigma_{eff} &= \sigma \Phi_C(C) \Phi_W(W_a), \\ \Phi_C(0) &= \Phi_W(0) = 1, \end{aligned} \tag{2}$$

where σ is the Stefan-Boltzmann constant. The "greenhouse" effect is described by the functions $\Phi_C(C)$ and $\Phi_W(W_a)$, which are monotonous decreasing with saturation levels Φ_C^∞ and Φ_W^∞ (they can be well approximated by Michaelis–Menten hyperbolas). The positive coefficient k_W stands for a hidden heat of evaporation. Finally, the "cloudiness parameter" n can be presented in the form

$$n = \begin{cases} 2q - 1, & \text{if } 0.5 \le q \le 1, \\ 0, & \text{if } 0 \le q < 0.5, \end{cases}$$

where q is a relative atmospheric humidity,

$$q = W_a / W_a^*,$$

$W_a^* = W_a^*(T)$ is a concentration of saturated water vapour depending on the temperature almost exponentially.

Albedo We consider a planet with vegetation and a "two-layer" atmosphere. Clouds and the underlying surface (vegetation and ocean can be taken into consideration) reflect the solar radiation, and the "greenhouse" gases transform it. A very realistic assumption is that

$$\alpha = \alpha(N, n, T, W_s),$$

where N is a vegetation density, and W_s is a soil moisture (i.e., a measure of water in soil). We set

$$\alpha = n\alpha^c + (1 - n)\alpha^{nc},$$

where α^c is the albedo of a sky covered with clouds, and α^{nc} is the albedo of a clear sky. Moreover,

$$\alpha^c = \alpha^n + (1 - \alpha^n)^2\alpha_{surf},$$

where α^n is the clouds' albedo and α_{surf} is the albedo of the underlying surface; the square arises because of the double reflection from the clouds and underlying surface. Since $\alpha^n \approx 0.5$, assume

$$\alpha^c = 0.5 + 0.25\alpha_{surf}.$$

Similarly, we have

$$\alpha^{nc} = \alpha_a + (1 - \alpha_a)^2\alpha_{surf},$$

where α_a is the albedo of the upper atmosphere without clouds ($\alpha_a \approx 0.1$). Then

$$\alpha^{nc} = 0.1 + 0.81\alpha_{surf}.$$

Finally,

$$\alpha = 0.1 + 0.4n + \alpha_{surf}(0.81 - 0.56n).$$

Let us specify α_{surf}. Considering a planet covered with snow and vegetation, without ocean, we can write

$$\alpha_{surf} = \{\lambda(N)\alpha_{bs} + [1 - \lambda(N)]\alpha_{veg}\}(1 - f_{sn}(T)) + \alpha_{sn}f_{sn}(T).$$

Here α_{sn} is the albedo of snow ($\alpha_{sn} \approx 0.7$) and α_{bs} is the albedo of bare soil,

$$\alpha_{bs} = \alpha_{bs}^0 f_{bs}(W_s/W_s^*),$$

where W_s^* is the moisture of saturated soil. The functions λ, f_{sn} and f_{bs} are monotonous decreasing, and $\lambda(N) \leq 1$, $f_{sn}(T) \leq 1$, $f_{bs}(W_s) \leq 1$.

Equation for vegetation A simple balance equation for N is

$$\frac{\partial N}{\partial t} = P - mN;$$

here P is a function of productivity (or annual net-production), $m = 1/\tau_N$, and τ_N is a residence time of carbon in the living biomass. In accordance with the Liebig principle the productivity (growth) function can be represented as

$$P = P_{max} \cdot g_T(T) \cdot g_C(C) \cdot g_N(N) \cdot g_W(W_s). \tag{3}$$

Here P_{max} is the maximal value of productivity corresponding to the abundance in resources and optimal values of all other parameters. The functions g_T and g_W are unimodular. In what follows, $[T_1, T_2]$ is an interval of temperature tolerance for vegetation, and $[W_s^1, W_s^2]$ is an interval of water tolerance; here W_s^1 is a point of wilting and W_s^2 is a point of full soil saturation. We have $g_T(T) = 0$ as $T \notin [T_1, T_2]$, and $g_W(W) = 0$ as $W \notin [W_s^1, W_s^2]$. Functions g_C and g_N are monotonous increasing with saturation.

Global carbon cycle In the present model we use a very primitive sub-model for the global carbon cycle. Let $D(x, t)$ be a density of carbon in soil (humus). We set

$$\begin{aligned} \frac{dC}{dt} &= -P + (1 - \varepsilon)mN + \delta(T, W_s)D + e(x, t), \\ \frac{dN}{dt} &= P - mN, \\ \frac{dD}{dt} &= \varepsilon mN - \delta(T, W_s)D. \end{aligned} \tag{4}$$

Here

$$\frac{d}{dt} = \frac{\partial}{\partial t} + V_\xi \nabla - D_\xi \nabla^2,$$

where V_ξ is the vector of a propagation velocity of carbon, and D_ξ is a measure of its motility (i.e., a matrix of diffusion coefficients) in corresponding media (atmosphere, biota, and soil). The function δ models a decomposition rate of soil carbon, and ε is a fraction of slowly decomposible carbon. We note that ε is in fact close to zero because carbon in litter is decomposed very fast (moreover, one should take into account the impact of respiration).

In agreement with the Liebig principle, one can assume a multiplicative form for δ:

$$\delta = \delta_{max} \cdot d_T(T) \cdot d_W(W_s),$$

where d_T and d_W are unimodular functions. The dependence of δ on T, or $d_T(T)$ is unimodular, since a leading process in carbon release from soil is microbial decomposition. The same argument applies for the dependence of δ on W_s, or $d_W(W_s)$. In dry soil corresponding to small W_s the rate δ is low, since the activity of microorganisms is low. In wet soil corresponding to large W_s the oxygen deficit inhibits the microorganisms' activity, and δ is again low.

The function $e(x, t)$ models the annual CO_2 anthropogenic emission. If this value is relatively small (for instance, $e(x, t) \ll P_{max}$), then we have a conservation law for carbon in the form

$$\int_{\Sigma} \{C(x,t) + N(x,t) + D(x,t)\} dx = A \approx \text{const.} \tag{5}$$

The domain of integration, Σ, is the whole surface of the planet. Later on we shall consider the value of A as a slowly changing bifurcation parameter.

Water (Hydrological) cycle A water balance equation (recall that we consider a planet without ocean) can be given in the form

$$\frac{dW_a}{dt} = -H + E + q_a(x,t),$$
$$\frac{dW_s}{dt} = H - E + q_s(x,t).$$

Here H and E are, respectively, annual precipitation and annual evaporation + evapotranspiration, and q_a and q_s model external sources and sinks. If q_a and q_s are relatively small, then we have a water conservation law in the form

$$\int_{\Sigma} \{W_a(x,t) + W_s(x,t)\} dx = B \approx \text{const.}$$

We can assume with a sufficient accuracy that precipitation is a linear function of the product $W_a \cdot n$, i.e.,

$$H = \nu_a(W_a \cdot n).$$

The total annual evaporation E can be represented in the form

$$E = \lambda(N)E_{bs} + (1 - \lambda(N))E_{veg},$$

where E_{bs} is the (physical) evaporation of bare soil,

$$\begin{aligned} E_{bs} &= f_e(T, W_a)W_s/W_s^*, \\ f_e &= W_a^*(T) - W_a. \end{aligned} \tag{6}$$

The functions $W_a^*(T)$ and $\lambda(N)$ have been characterized earlier. A good assumption on evapotranspiration by vegetation is that it is proportional to the productivity, i.e.,

$$E_{veg} = \nu_v P.$$

Thus, we derived a complete closed system of differential equations for a description of the "climate + biosphere" dynamics.

3 Zero-dimensional Global Carbon Cycle Model

Let us analyze a zero-dimensional (in space) global carbon cycle model without a water cycle in more detail. Setting $n = 0$ and referring to (2) we rewrite the climate (temperature) equation (1) as

$$k\dot{T} = S(1 - \alpha) - \sigma\Phi_C(C)T^4.$$

The equation for biota (vegetation) has the form (see (4))

$$\frac{dN}{dt} = P(C, N, T) - mN.$$

Recall that P is the annual net-production of vegetation, and m is the inverse for τ_N, the residence time of carbon in biota. According to (3) $P = P_{\max} \cdot g_T(T) \cdot g_C(C) \cdot g_N(N)$. Since $C = A - N$, (see (5)) then the product $g_C(C) \cdot g_N(N)$ can be represented as $G_N(N)$. Thus,

$$P = P_{max} \cdot g_T(T) \cdot G_N(N).$$

The function G_N defined on the interval $N \in [0, A]$ is unimodular, and satisfies $G_N(0) = G_N(A) = 0$. Setting $\delta = 0$ and $\varepsilon = 0$, we specify the equation for atmospheric carbon (see (4)) into

$$\frac{dC}{dt} = -P(C, N, T) + mN + e(t), \tag{7}$$

where $e(t)$ is the annual anthropogenic emission. The total carbon conservation law (5) takes the form

$$C(t) + N(t) = A_0 + \int_{t_0}^{t} e(\tau)d\tau = A(t).$$

Using this equality to exclude the variable $C(t)$ from the equations, we get

$$\frac{dT}{dt} = \frac{1}{k}\left[\Psi(N) - \sigma\varphi(N)T^4\right], \tag{8}$$

$$\frac{dN}{dt} = P_{max} \cdot g_T(T) \cdot G_N(N) - mN, \tag{9}$$

where

$$\Psi(N) = S(1 - \alpha(N))$$

and

$$\varphi(N) = \Phi_C(A - N). \tag{10}$$

The system (8), (9) is a simplest model for the biosphere of our hypothetical planet.

Equilibria For $A(t) = A = \text{const}$ the equilibrium points (T^*, N^*) of the system (8), (9) are determined by

$$T^* = \left(\frac{\Psi(N^*)}{\sigma\varphi(N^*)}\right)^{1/4}, \tag{11}$$

$$g_T(T^*) \cdot G_N(N^*) = \frac{m}{P_{max}} N^*.$$

Noticing that $G_N(0) = 0$, we represent G_N in the form $G_N(N) = f(N) \cdot N$, where

$$\lim_{N \to 0} f(N) = f(0) < \infty, \quad f'(N) < 0.$$

Then

$$N^* = 0, \quad T_0^* = \left(\frac{\Psi(0)}{\sigma\varphi(0)}\right)^{1/4}$$

is an equilibrium. Since $N^* = 0$, this equilibrium corresponds to a "naked" planet, i.e., a desert with no vegetation and all carbon accumulated in the atmosphere. All other equilibria are the solutions of the system of equations (11) and

$$g_T(T^*) = \frac{m}{P_{max}} \cdot \frac{1}{f(N^*)}.$$

For these equilibria $g_T(T^*) > 0$, i.e., T^* lies in $[T_1, T_2]$, the interval of temperature tolerance for vegetation. Applying the Dulac criterion, we can prove that the system (8), (9) has no closed trajectories.

Stability analysis After the linearization of the system (8), (9) in a vicinity of an equilibrium point (N^*, T^*), we get the Jacobi matrix

$$J = \begin{pmatrix} -4\frac{\Psi(N^*)}{kT^*} & \left[\ln\frac{\Psi(N)}{\varphi(N)}\right]'_{N^*} \frac{\Psi(N^*)}{k} \\ P_{max}N^*f(N^*)g'_T(T)_{T^*} & P_{max}g_T(T^*)[f(N^*) + N^*f'(N^*)] - m \end{pmatrix}.$$

Let us consider the point $(N^* = 0, T_0^*)$. We have

$$J_0 = \begin{pmatrix} -4\frac{\Psi(0)}{kT_0^*} & \left[\ln\frac{\Psi(N)}{\varphi(N)}\right]'_0 \frac{\Psi(0)}{k} \\ 0 & P_{max}g_T(T_0^*)f(0) - m \end{pmatrix}.$$

The corresponding eigenvalues are

$$\lambda_1 = -4\frac{\Psi(N^*)}{kT^*} < 0, \quad \lambda_2 = P_{max}g_T(T^*)f(0) - m.$$

This equilibrium is a stable node if

$$g_T(T_0^*) < \frac{m}{P_{max}f(0)}$$

and an unstable saddle, if

$$g_T(T_0^*) > \frac{m}{P_{max}f(0)}.$$

Since $g_T(T) = 0$ for $T \notin (T_1, T_2)$, for any equilibrium $(N^* = 0,\ T_0^* \notin (T_1, T_2))$ we have $\lambda_2 = -m$, hence every such equilibrium is stable.

Let $N^* \neq 0$. Then

$$J = \begin{pmatrix} -4\frac{\Psi(N^*)}{kT^*} & \left[\ln\frac{\Psi(N)}{\varphi(N)}\right]'_{N^*}\frac{\Psi(N^*)}{k} \\ mN^*[\ln g_T(T)]'_{T^*} & mN^*[\ln f(N)]'_{N^*} \end{pmatrix},$$

and the corresponding eigenvalues are

$$\lambda_{1,2} = -\frac{1}{2}\left(\frac{4\Psi(N^*)}{kT^*} - mN^*(\ln f(N))'_{N^*}\right)$$
$$\pm\frac{1}{2}\sqrt{\left[\frac{4\Psi(N^*)}{kT^*} + mN^*(\ln f(N))'_{N^*}\right]^2 + \frac{4\Psi(N^*)mN^*}{k}(\ln g_T)'_{T^*}\left(\ln\frac{\Psi(N)}{\varphi(N)}\right)'_{N^*}}.$$

Since $(\ln f(N))' < 0$ for any $N \leq A$, we have

$$\frac{4\Psi(N)}{kT} - mN(\ln f(N))' > 0,$$

and the considered equilibrium, with $N^* \neq 0$, cannot be an unstable node. This equilibrium is a saddle point if

$$F(N^*, T^*) = \left(\ln\frac{\Psi(N)}{\varphi(N)}\right)'_{N^*}(\ln g_T(T))'_{T^*} + 4(\ln f(N))'_{N^*}(\ln T)'_{T^*} > 0.$$

If $F(N^*, T^*) < 0$, we have a stable node or a stable focus; moreover, for the latter it holds that

$$\left[\frac{4\Psi(N^*)}{kT^*} + mN^*(\ln f(N))'_{N^*}\right]^2 + \frac{4\Psi(N^*)mN^*}{k}(\ln g_T(T))'_{T^*}\left(\ln\frac{\Psi(N)}{\varphi(N)}\right)'_{N^*} < 0.$$

4 Parametrization

In order to make our investigation simpler and more visual, we use specific parametric representations for functions $\Psi(N)$, $\varphi(N)$, and $f(N)$.

1. Let

$$\Phi_C(C) = 1 - \frac{(1-\varphi_\infty)C}{k_c + C}$$

where $0 < \varphi_\infty < 1$. Then (see (10))

$$\varphi(N) = \frac{k_c + \varphi_\infty(A - N)}{k_c + A - N}. \tag{12}$$

A good representation for $\Psi(N)$ is

$$\Psi(N) = s_1 + \frac{(s_2 - s_1)N}{k_\alpha + N} = \frac{s_1 k_\alpha + s_2 N}{k_\alpha + N}.$$

Then

$$\Phi(N) = \frac{\Psi(N)}{\varphi(N)} = \frac{(s_1 k_\alpha + s_2 N)(\chi - N)}{(k_\alpha + N)(k_c + \varphi_\infty A - \varphi_\infty N)},$$

where $\chi = k_c + A$, or

$$\Phi(N) = \frac{s_2}{\varphi_\infty} \cdot \frac{(a_1 + N)(\chi - N)}{(k_\alpha + N)(\chi_1 - N)},$$

where $a_1 = s_1 k_\alpha / s_2$, $\chi = k_c + A$, and $\chi_1 = \frac{k_c}{\varphi_\infty} + A$. It is obvious that $\chi_1 > \chi > A$, and $a_1 < k_\alpha$.

2. Let

$$g_T(T) = \begin{cases} \frac{4}{\Delta T^2}(T - T_1)(T_2 - T), & \text{if } T \in [T_1, T_2] \\ 0, & \text{if } T \notin [T_1, T_2] \end{cases},$$

where ΔT is the length of the tolerance interval for vegetation, i.e., $\Delta T = T_2 - T_1$. Note that $T_{opt} = (T_1 + T_2)/2$ is a maximizer for g_T, and $g_T(T_{opt}) = 1$.

3. To introduce a parametrization for $f(N)$, we recall that $f(N) = G_N(N)/N$. Let $G_N(N)$ be described by a parabola,

$$G_N(N) = \frac{4}{A^2} N(A - N),$$

and, respectively,

$$f(N) = \frac{4(A - N)}{A^2}.$$

Using this parametrization we can investigate the evolution of the phase plane with the increase of the bifurcation parameter A.

A "naked" planet Let us consider $(N^* = 0, T_0^*)$, the equilibrium of a "naked" planet, in detail. If $T_0^* \notin (T_1, T_2)$, i.e. the equilibrium temperature lies beyond the tolerance interval for photosynthesis, then this equilibrium is stable. If $T_0^* \in (T_1, T_2)$ then this equilibrium is stable if

$$g_T(T_0^*) < \frac{m}{P_{max} f(0)} = \frac{m}{4P_{max}} A,$$

or

$$g_T\left(\sqrt[4]{\frac{1}{\sigma}\Phi(0)}\right) < \frac{m}{4P_{max}}A. \tag{13}$$

Since

$$\Phi(0) = \frac{S(1-\alpha(0))}{\varphi(A)},$$

the left-hand side in (13) depends only on A, a total amount of carbon in the system. The right-hand side in (13) depends on A and the characteristics of vegetation. If we assume that T^*, the equilibrium temperature for a "naked" planet, and A, the total amount of carbon, are fixed, then, changing the vegetation characteristics (in an evolutionary way), we pass from a stable "naked planet" equilibrium to an unstable one characterized by

$$g_T\left(\sqrt[4]{\frac{1}{\sigma}\Phi(0)}\right) > \frac{m}{4P_{max}}A. \tag{14}$$

The latter inequality and a shape of the function g_T show that the condition that the equilibrium temperature belongs to the temperature tolerance interval for photosynthesis ($T^* \in [T_1, T_2]$) is not sufficient for reaching an unstable "naked planet" equilibrium, i.e., an "origin of life". A "naked planet" equilibrium becomes unstable and "life" may occur whenever $T^* \in (T_1', T_2')$, where $(T_1', T_2') \subset [T_1, T_2]$. The interval (T_1', T_2') depends on $\tau_N = 1/m$, a residence time of carbon in the biota, and its maximum productivity, P_{max}. We call (T_1', T_2') the vegetation tolerance interval. Let us vary the parameter

$$\beta = \frac{m}{4P_{max}} = \frac{1}{4P_{max}\tau_N}.$$

Observing (14), we see that the increase of β corresponding to the decrease of a carbon residence time in the biota or a decrease of a maximum productivity of photosynthesis leads to a reduction of the vegetation tolerance interval, and, vice versa, the decrease of β, or the increase of either a residence time or a maximal productivity leads to an extension of the vegetation tolerance interval. Since $g_T(T) \leq 1$, we find that whenever $\beta A > 1$, a "naked planet" equilibrium is stable for any T^*, and "life" cannot arise in a vicinity of this equilibrium. In other words, there is some critical combination of P_{max}, m (or τ_N), and A: a "naked" planet equilibrium can be unstable only if

$$\frac{mA}{4P_{max}} < 1,$$

or

$$A < 4P_{max}\tau_N.$$

In fact there are two bifurcation parameters, β and A.

Observing (11) and (12), we see that

$$T^* = T^*(A) = \left(\frac{\Phi(0, A)}{\sigma}\right)^{1/4} = \left(\frac{S(1 - \alpha(0))}{\sigma\varphi(0, A)}\right)^{1/4}.$$

Since $\varphi_A \to \varphi_\infty = \text{const}$ as $A \to \infty$, then $T_0^*(A) \to (T_0^*)_{max} = \text{const}$ as $A \to \infty$. Since $\varphi(0, A) = 1$ for $A = 0$, we have $T_0^*(A) \to (T_0^*)_{min}$ as $A \to 0$, where

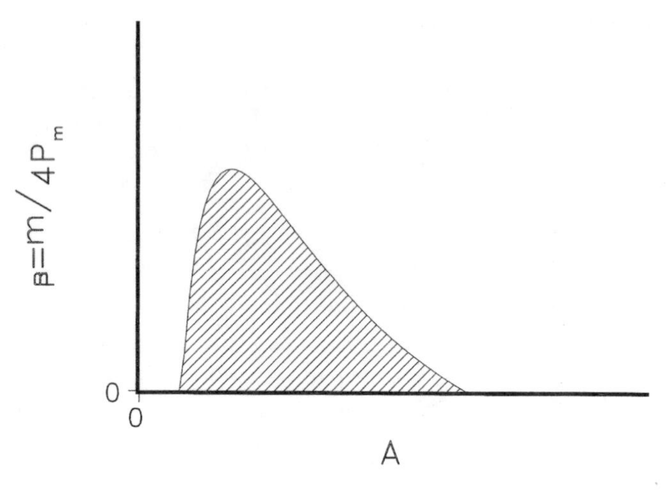

Figure 2 Stability border for $N^* = 0$ in the $\{\beta, A\}$ domain: $g_T(T_0^*) = \beta A$, $T_0^* = T_0^*(A)$. The shaded area indicates the area of instability of $N^* = 0$ where "life" can arise in the vicinity of the "naked" equilibrium.

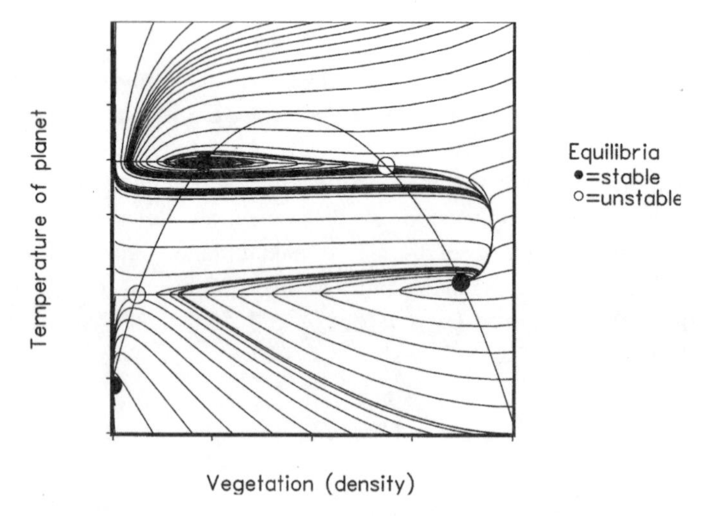

Figure 3 Equilibria of the climate–vegetation model.

$$(T_0^*)_{min} = \left(\frac{S(1 - \alpha(0))}{\sigma} \right)^{1/4}.$$

Let us assume that $(T_0^*)_{min} < T_1$, and $(T_0^*)_{max} > T_2$. Then we have the following stability diagram (see Fig. 2).

Fig. 3 shows schematically a location of equilibria of the climate–vegetation model. Three of the five equilibria are stable.

5 Conclusion

The presented climate–vegetation model involving a carbon cycle submodel shows some interesting features. The performed stability analysis of the system's equilibria gives us up to three stable points. Depending on an initial condition, a stable "cold desert" with no vegetation, a "cold planet" with a large amount of vegetation, and a "hot planet" with a small amount of vegetation (compared to a "cold planet") can exist. The initiation of life on our virtual planet depends on the value of two bifurcation parameters, the total amount of carbon A in the system and a combination of biotic characteristics β of the vegetation.

The stability properties of different equilibrium states can change dramatically due to the increase of the anthropogenic emissions. There is a possibility that a "cold planet" becomes unstable and a new "hot planet" equilibrium is realized. Such a transition forces a drastic decrease in the vegetation of the planet.

References

[1] Kostitzin, V.A. (1935) *L'Evolution de l' Atmosphere: Circulation Organique, Epoques Glaciaires*. Hermann, Paris.
[2] Schellnhuber, H.J. and von Bloh, W. (1993) Homöostase und Katastrophe, in: Schellnhuber, H.J. and Sterr, H. (Editors), *Klimaänderung und Küste*. Springer Verlag, Berlin, pp. 11–27.
[3] Svirezhev, Yu. (1994) Simple model of interaction between climate and vegetation: virtual biospheres. IIASA Seminar, Laxenburg.
[4] Vernadsky, V.I. (1926) *The Biosphere*. Nauchtechizdat, Leningrad.
[5] Watson, A.J. and Lovelock, J.E. (1983) Biological homeostasis of the global environment: the parable of Daisyworld. *Tellus* 35B, 284–289.

17 ECOLOGICAL MODELING AND COUPLED EXPONENTIAL MAPS

NARAYANAN RAJU[1] and FIRDAUS E. UDWADIA[2]

[1]*Applied Simulation Technology, 1641 N. First St. Suite 170, San Jose, CA 95112, USA*
[2]*Department of Mechanical Engineering and Department of Aerospace Engineering, University of Southern California, Los Angeles, CA, USA*

1 Introduction

In the past decade alone more than 5000 articles have been published on the chaotic behavior of dynamical systems [11]. Such studies cover a wide range of subjects including ecology, economics, biology, chemistry, engineering and other physical sciences. In nearly every nook and corner of science and technology, chaos has been reported. The number of such reports is furthermore increasing at a rapid rate. But common experience indicates that most physical systems, even nonlinear ones, seldom behave erratically and are usually not highly sensitive to initial conditions.

If most nonlinear systems exhibit chaos, why then do we see so much stability in the real world? A natural question that arises is: How does stability originate despite the fact that the nonlinear "units" from which a system may be made up are chaotic, and that this chaotic behavior appears to be ubiquitous. The aim of this paper then is to increase our understanding of the cause and nature of the stability present in physical systems.

There may be many reasons for the observed stability of physical systems present amidst all this reported chaos. In a given system there are several parameters and chaos may occur for only certain ranges of these parameters. If the probability that the system attains these ranges of parameters is small, then the probability of this system being chaotic is also small. However, estimating such probabilities may be difficult.

Further, in reality, most physical systems are coupled. To render them easy to study, many of the couplings present in physical systems are either eliminated, approximated or assumed away. Therefore, another possibility is that a chaotic system may become stable when coupled with other chaotic systems. Cast differently, can adding a "jug" of chaos to another "jug" of chaos lead to stability? For example, if one group (or species) that displays chaotic growth (in isolation) is coupled with another chaotically evolving group, could the resulting system exhibit orderly dynamical behavior? Could, for example, migratory coupling of two chaotically evolving insect populations stabilize these populations? In short, can the collective behavior of a system transcend that of its chaotic components? Our investigations demonstrate the possibility of such stabilization (Fig. 1).

The nonlinear model used here to study the role of coupling in stabilizing and synchronizing nonlinear systems is derived from population dynamics, though several of our results appear valid for general nonlinear maps.

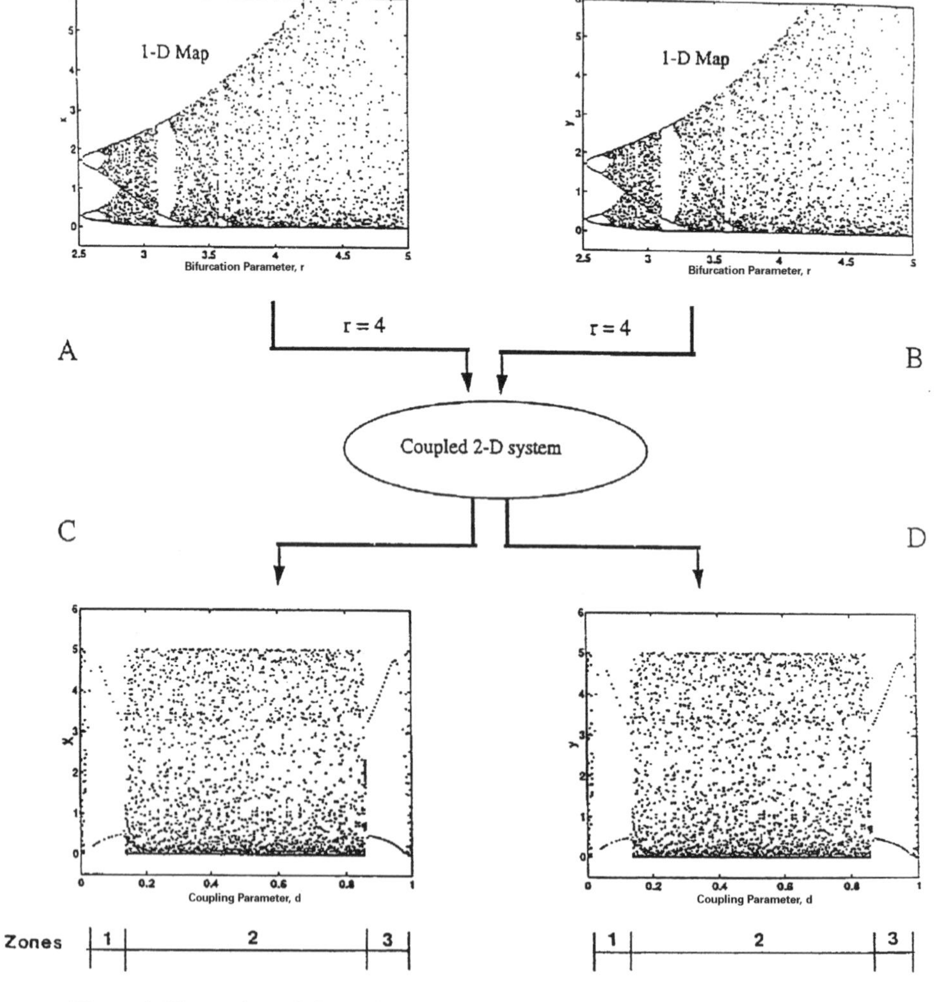

Figure 1 Illustration of the stability added by coupling two chaotic 1-D maps.

2 The Population Model

We analyze the effect of migration by studying the interactive dynamics of two subcolonies of a single species. The growth of a single subcolony is often modeled as a difference equation of the type

$$x_{n+1} = p(x_n) \tag{1}$$

Here x_n is the population size at time t_n, and the map p gives the size of the population at the next update in time, t_{n+1}. Exponential maps, wherein the function p is defined as

$$p(x) = xe^{r(1-x)} \tag{2}$$

are used extensively in the biological literature [8] to describe a population with a propensity to grow exponentially at low population densities and a tendency to decrease at high density. The nature of the nonlinear behavior is regulated by the growth parameter r. Ricker [8], based on empirical data, has postulated that exponential maps govern the population of a prey species if the predators, at any given abundance, consume a fixed fraction of the prey species, as though they were captured through random encounters. The same model also governs the population dynamics of a single species of organisms which is regulated by an epidemic disease at high population densities.

The system described by equations (1) and (2) shows complicated dynamics. May [6] provides some of the salient features of this map as the parameter r is gradually varied. Starting from a low value of the growth parameter r, the fixed point ($x = 1$) of the map becomes unstable when r reaches a value of 2 [Fig. 1]. The point of accumulation of cycles of period 2^n occurs when $r \simeq 2.6924$, and the 3-period cycle appears at $r \simeq 3.1024$. Beyond this value of r, the system exhibits chaotic behavior interspersed by thin periodic bands. The chaotic or periodic behavior of the system is reflected in the positive or negative values for the Lyapunov exponent plotted in Fig. 1.

In this paper we consider two interacting populations (colonies) of biological organisms, each of whose population dynamics is described by equations (1) and (2). The interaction between these colonies may be thought of as being brought about by migration between the two populations. In this paper we concentrate on the situation where the growth rates are identical.

The similar case of coupled logistic maps has been analyzed by Gyllenberg *et al.* [3] to study the effect of migration on population dynamics. They have presented a detailed analysis of fixed points and two-periodic orbits. Coupled logistic equations with different forms of coupling have been investigated by, among others, Chowdhury and Chowdhury [1] and Kaneko [4]. These studies have focussed on studying the periodic trajectories. We, on the other hand, take two or more chaotic units and analyze the effect of coupling in stabilizing the otherwise chaotic dynamics.

Our aim is to investigate the stable and chaotic dynamics of the coupled system. The effect of coupling on the system dynamics is investigated in detail. *It is demonstrated that coupling has a stabilizing effect.*

We consider two interacting populations denoted by x and y of two colonies of the same species of organisms. Assuming that the populations grow as per the exponential map, the population in the first location at the end of one time period (say, one year) is $x\exp[r_x(1-x)]$ while that at the second location is $y\exp[r_y(1-y)]$. Further, if only fraction d of the population remains at the same location while the rest emigrate to the other location, we can arrive at the following equation governing the two populations x and y

$$x_{n+1} = dx_n \exp[r_x(1-x_n)] + (1-d)y_n \exp[r_y(1-y_n)] \tag{3}$$

$$y_{n+1} = (1-d)x_n \exp[r_x(1-x_n)] + dy_n \exp[r_y(1-y_n)]$$

where as usual the subscript n denotes the time t_n. The growth rates for the two subcolonies are given by r_x and r_y. The parameter d describes the extent of coupling between the two populations. When $d=1$, the two populations evolve in time independently of each other; we have two uncoupled maps each describing the population count of one colony at any time t_n. On the other hand, when $d=0$, the populations are most intensely coupled; the entire population produced at one location migrates to the other. When $0 \leq d < 1$, the two colonies interact with each other. The coupling used here is symmetric, i.e., the percentage migrating from one location is the same as that migrating from the other. Such a symmetry reduces the number of parameters in the problem. We shall sometimes denote the mapping described by equation (3), for brevity, as

$$(x_{n+1}, y_{n+1}) = M(r_x, r_y, d) \circ (x_n, y_n), \tag{4}$$

to indicate explicitly the dependence on the triplet of parameters (r_x, r_y, d).

We also investigate the effect of increasing the number of habitats that are coupled. We assume that more than two habitats can be present on a ring and adjacent habitats are coupled through migration. We let a fraction d stay at the location they were born and the rest $(1-d)$ migrate to the two adjacent habitats on either side in equal proportions i.e., we let a fraction $(1-d)/2$ migrate to each of the two adjacent habitats. In this way we can simulate the population dynamics of three or more habitats on a ring.

3 Results

When only two habitats are coupled we can derive several global results. We first present these results and numerical results are presented in Section 3.2.

3.1 Analytical Results

Result 1 Consider the orbits of the map (4) for a specific set of parameters $r_x = \hat{r}_x$, $r_y = \hat{r}_y$, and $d = \hat{d}$. For each orbit of the map (4) described by $\{(\hat{x}_n, \hat{y}_n) \mid n = 0, 1, 2, \ldots\}$ for a given triplet $(\hat{r}_x, \hat{r}_y, \hat{d})$ of parameter values, there corresponds an orbit described by $\{(\hat{y}_n, \hat{x}_n) \mid n = 0, 1, 2, \ldots\}$ for the parameter triplet $(\hat{r}_y, \hat{r}_x, \hat{d})$.

Proof This result is obvious from equations (3) since the interchanges

$$\begin{Bmatrix} x_n \to y_n \\ y_n \to x_n \\ \hat{r}_x \to \hat{r}_y \\ \hat{r}_y \to \hat{r}_x \end{Bmatrix} \Longrightarrow \begin{Bmatrix} x_{n+1} \to y_{n+1} \\ y_{n+1} \to x_{n+1} \end{Bmatrix}$$

□

Result 2 When $r_x = r_y = \hat{r}$, for each orbit $\{(\hat{x}_n, \hat{y}_n) \mid n = 0, 1, 2, \ldots\}$, corresponding to the parameter values $(\hat{r}, \hat{d})$, corresponds an orbit $\{(\hat{y}_n, \hat{x}_n) \mid n = 0, 1, 2, \ldots\}$.

Proof The result follows directly from Result 1 by setting $r_x = r_y = \hat{r}$. □

Corollary 1 When $r_x = r_y = \hat{r}$, for each n-periodic orbit of the map (3) described by $\{(\hat{x}_n, \hat{y}_n) \mid n = 0, 1, 2, \ldots\}$ corresponding to a set of parameters $(\hat{r}, \hat{d})$, there exists another n-periodic orbit described by $\{(\hat{y}_n, \hat{x}_n) \mid n = 0, 1, 2, \ldots\}$.

Proof The result is a special case of Result 2 when the orbits are periodic. □

Result 3 Consider the orbit $\{(\hat{x}_n, \hat{y}_n) \mid n = 0, 1, 2, \ldots\}$ corresponding to a certain value of the parameter $d = 1/2 - d_0$, $0 \leq d_0 \leq 1/2$, with $r_x = r_y = \hat{r}$. For each such orbit, there corresponds an orbit given by $\{(\hat{x}_0, \hat{y}_0), (\hat{y}_1, \hat{x}_1), (\hat{x}_2, \hat{y}_2), (\hat{y}_3, \hat{x}_3), \ldots\}$, corresponding to the parameter $d = 1/2 + d_0$, with $r_x = r_y = \hat{r}$, i.e., each alternate point of the second orbit has its x- and y-coordinate switched with respect to the corresponding point of the first orbit.

Proof The result follows from the observation that for an initial point $(\hat{x}_n, \hat{y}_n)$, if

$$(\hat{x}_{n+1}, \hat{y}_{n+1}) = M(\hat{r}, \hat{r}, 1/2 - d_0) \circ (\hat{x}_n, \hat{y}_n), \tag{5}$$

then,

$$(\tilde{x}_{n+1}, \tilde{y}_{n+1}) = M(\hat{r}, \hat{r}, 1/2 + d_0) \circ (\hat{x}_n, \hat{y}_n) = (\hat{y}_{n+1}, \hat{x}_{n+1}) \tag{6}$$

Furthermore denoting

$$(\hat{x}_{n+2}, \hat{y}_{n+2}) = M(\hat{r}, \hat{r}, 1/2 - d_0) \circ (\hat{x}_{n+1}, \hat{y}_{n+1}), \tag{7}$$

we find that

$$(\tilde{x}_{n+2}, \tilde{y}_{n+2}) = M(\hat{r}, \hat{r}, 1/2 + d_0) \circ (\tilde{x}_{n+1}, \tilde{y}_{n+1}) = (\hat{x}_{n+2}, \hat{y}_{n+2}), \tag{8}$$

and hence the result. □

Corollary 2 If the map $M(\hat{r}, \hat{r}, 1/2 - d_0)$, $0 \leq d_0 \leq 1/2$ has a 2n-period orbit starting from some (x_0, y_0), $n = 1, 2, \ldots$, then the map $M(\hat{r}, \hat{r}, 1/2 + d_0)$ must have a 2n-period starting from the same point (x_0, y_0), $n = 1, 2, \ldots$.

Proof This again is obvious from Result 3. □

Corollary 3 If the map $M(\hat{r}, \hat{r}, 1/2 - d_0)$, $0 \leq d_0 \leq 1/2$ has a $(2n-1)$-period orbit starting some (x_0, y_0), $n = 1, 2, \ldots$, with $x_0 \neq y_0$, then the map $M(\hat{r}, \hat{r}, 1/2 + d_0)$ must have a $2(2n-1)$-period starting from the same point (x_0, y_0), $n = 1, 2, \ldots$.

Proof This again is obvious from Result 3. □

As we shall see from numerical simulations, this property of the map yields a sort of quasi-symmetry to the bifurcation plots of the orbits with respect to the parameter d.

Result 4 For $r_x = r_y = \hat{r}$ the orbit of map (4) starting with (x_0, x_0) will always consist of points of the form (x_n, x_n). In other words, orbits which begin on the diagonal in the (x, y) phase space lie entirely on the diagonal.

Proof This follows from the observation that

$$(x_{n+1}, x_{n+1}) = M(\hat{r}, \hat{r}, d) \circ (x_n, x_n) \tag{9}$$

□

The above result indicates that for orbits that begin on the diagonal in the phase space, each state of the dynamical system behaves as if it were governed by a single exponential map of the form of equation (2).

Result 5 For $d = 1/2$ the orbit of the map (4) starting with (x_0, y_0) will consist of points of the form (x_n, x_n) for all values of r_x and r_y after the first iteration.

Proof This result is obvious from the observation that the map (4) redistributes the total population equally among the two locations. □

We now consider the fixed points of the map (4). It is easily seen that the points (0,0) and (1,1) are fixed points.

Result 6 The fixed point $(0, 0)$ of the map $M(r_x, r_y, d)$ is unstable for all $r_x, r_y > 0$ and $0 \leq d \leq 1$.

Proof The Jacobian matrix of the map $M(r_x, r_y, d)$ evaluated at (x, y) is given by

$$J(x, y) = \begin{bmatrix} d(1 - xr_x)\exp[r_x(1-x)] & (1-d)(1 - yr_y)\exp[r_y(1-y)] \\ (1-d)(1 - xr_x)\exp[r_x(1-x)] & d(1 - yr_y)\exp[r_y(1-y)] \end{bmatrix} \tag{10}$$

Hence,

$$J(0, 0) = \begin{bmatrix} d\exp(r_x) & (1-d)\exp(r_y) \\ (1-d)\exp(r_x) & d\exp(r_y) \end{bmatrix}, \tag{11}$$

and thus for (0,0) to be stable we require, using the usual determinant-trace criterion of the Jacobian matrix, that

$$(2d-1)\exp(r_x+r_y)<1, \tag{12}$$

and

$$(2d-1)\exp(r_x+r_y)\pm d[\exp(r_x)+\exp(r_y)]+1>0 \tag{13}$$

For condition (12) to be true

$$d<\frac{1}{2}+\frac{1}{2\exp(r_x+r_y)}$$

Let

$$d=\frac{1}{2}+\frac{1}{2\exp(r_x+r_y)}-\epsilon$$

To satisfy condition (13),

$$2-2\epsilon\exp(r_x+r_y)-[\exp(r_x)+\exp(r_y)]>0$$

which cannot be satisfied for any positive values for r_x and r_y. Hence the result. □

Result 7 The fixed point (1,1) is stable when

$$(1-r_x)(1-r_y)(2d-1)<1, \tag{14}$$

and

$$(1-r_x)(1-r_y)(2d-1)\pm d(2-r_x-r_y)+1>0. \tag{15}$$

Proof the result is obvious by evaluating the Jacobian matrix given by equation (10) at (1,1). □

When $r_x=r_y=\hat{r}$, equations (14) and (15) simplify and it can be shown that for $\hat{r}<2$, the fixed point (1,1) is stable for all $0\leq d\leq 1$.

Besides these two fixed points which occur for all values of the triplet (r_x, r_y, d), there are other fixed points whose coordinates depend on specific values of the triplet.

Result 8 A fixed point (x_i, y_i) of the map $M(r_x, r_y, d)$ must satisfy the condition

$$\frac{r_x-\ln(1-dq)+\ln[d+(1-2d)]}{r_x[r_y-\ln(q)]}=\frac{d+(1-2d)q}{r_y(1-d)} \tag{16}$$

$$q=e^{r_y(1-y_i)} \quad \text{and} \quad x_i=y_i\frac{d+(1-2d)q}{1-d}$$

When $r_x=r_y=\hat{r}$, this equation simplifies to

$$\frac{\hat{r}-\ln(1-dq)+\ln[d+(1-2d)]}{\hat{r}[\hat{r}-\ln(q)]}=\frac{d+(1-2d)q}{\hat{r}(1-d)} \tag{17}$$

Proof The result follows directly from the definition of the map. □

Next we consider some properties of the Lyapunov Exponents λ_i,

$$\lambda_i = \lim_{n\to\infty} \frac{\log_e[|\mu_i|]}{n} \tag{18}$$

where μ_i are the eigenvalues of the product of the Jacobian matrices at every iteration.

Result 9 The Lyapunov exponents of the map $M(\hat{r}, \hat{r}, 1/2 - d_0)$, $0 \leq d_0 \leq 1/2$ starting from some (x_0, y_0) are the same as those of the map $M(\hat{r}, \hat{r}, 1/2 + d_0)$ starting from the same point (x_0, y_0).

Proof Let

$$M(\hat{r}, \hat{r}, 1/2 - d_0) \circ (x_n, y_n) = (x_{n+1}, y_{n+1}) \tag{19}$$

By Result 3,

$$M(\hat{r}, \hat{r}, 1/2 + d_0) \circ (x_n, y_n) = (\tilde{x}_{n+1}, \tilde{y}_{n+1}) = (y_{n+1}, x_{n+1}) \tag{20}$$

and

$$M(\hat{r}, \hat{r}, 1/2 - d_0) \circ (x_{n+1}, y_{n+1}) = M(\hat{r}, \hat{r}, 1/2 + d_0) \circ (\tilde{x}_{n+1}, \tilde{y}_{n+1}) = (x_{n+2}, y_{n+2}) \tag{21}$$

By Eq. (10), the Jacobian matrix for these two systems at every iteration is of the form

$$J_n[M(\hat{r}, \hat{r}, 1/2 - d_0)]_{(x_n, y_n)} = \begin{bmatrix} a_{11} & a_{12} \\ a_{21} & a_{22} \end{bmatrix} \tag{22}$$

$$\tilde{J}_n[M(\hat{r}, \hat{r}, 1/2 + d_0)]_{(\tilde{x}_n, \tilde{y}_n)} = \tilde{J}_n[M(\hat{r}, \hat{r}, 1/2 + d_0)]_{(x_n, y_n)} = \begin{bmatrix} a_{21} & a_{22} \\ a_{11} & a_{12} \end{bmatrix} \tag{23}$$

$$J_{n+1}[M(\hat{r}, \hat{r}, 1/2 - d_0)]_{(x_{n+1}, y_{n+1})} = \begin{bmatrix} b_{11} & b_{12} \\ b_{21} & b_{22} \end{bmatrix} \tag{24}$$

$$\tilde{J}_{n+1}[M(\hat{r}, \hat{r}, 1/2 + d_0)]_{(\tilde{x}_{n+1}, \tilde{y}_{n+1})} = \tilde{J}_{n+1}[M(\hat{r}, \hat{r}, 1/2 + d_0)]_{(y_{n+1}, x_{n+1})} = \begin{bmatrix} b_{12} & b_{11} \\ b_{22} & b_{21} \end{bmatrix} \tag{25}$$

Hence, it is seen that

$$J_{n+1} J_n = \tilde{J}_{n+1} \tilde{J}_n \tag{26}$$

Therefore the eigenvalues and the Lyapunov exponents are symmetric about $d = 1/2$. □

Result 10 When $d = 1/2$, one Lyapunov exponent tends to $-\infty$.

Proof On setting $d = 1/2$ in Eq. (10) the determinant vanishes and hence the result. □

We would like to reiterate that results 2 to 5, 9, 10, and their corollaries are independent of the functional form of the maps chosen. We next present some numerical results for the case when the growth rates (r_x, r_y etc.) in all the habitats are identical.

3.2 *Numerical Results*

We begin with a description of the coupled dynamics when only two adjacent habitats are coupled. As illustrated in Fig. 1 we take two populations which are evolving chaotically with a growth rate of $r = 4$. Their dynamics is studied by means of bifurcation plots. To arrive at the bifurcation plots of the coupled system we first apply the map 5000 times iteratively on an initial condition for a particular value of the growth parameter. The species-count at both locations denoted by $y1$ and $y2$ are then obtained by iterating another hundred times. This procedure is repeated for several values of d in the range zero to unity. The last hundred iterations are plotted in the figures.

It can readily be seen that for two ranges of the coupling parameter ($0.05 < d < 0.18$ and $0.82 < d < 0.95$) completely stable behavior is achieved. These two ranges are furthermore symmetric about $d = 1/2$ in agreement with Result 3 and its corollaries. The other symmetry properties present in the dynamics are outlined in a more comprehensive treatment elsewhere [10]. A large portion of the range $0.18 < d < 0.95$ furthermore shows chaotic though synchronized motion [10]. Similar synchronized motion has been observed by [2], [9], and [7]. Although not shown here, we have analyzed this system for several values of the growth parameter r for which the uncoupled dynamics is chaotic (say, $r = 3.25, 3.5, 3.8\ldots$) and found that the coupled system exhibits a similar stabilization process. We also verified that the dynamical behavior described in Fig. 1 is typical for several sets of the initial population sizes at each of these locations [10].

Here we analyze the effect of the number of coupled systems on the stability engendered. As discussed earlier, when more than two units are coupled, we place them on a ring and assume that a fraction d ($0 \leq d \leq 1$) of the species remains at the same location while the rest ($1 - d$) migrate to the two adjacent habitats in equal proportions. Thus there is only one coupling parameter d even if several habitats are coupled. Again d can vary from zero to unity. The other parameter in the system is the growth rate r, which is assumed to be the same for all sets of habitats. For a given growth rate r, we then study how the ranges of d for which stability is engendered varies with the number of units coupled, n.

We choose the value of the growth parameter such that the uncoupled dynamics are chaotic, say $r = 3.25$. Plotted in Fig. 2 are the bifurcation diagrams of the population at the first site for different values of n. The bifurcation diagrams for other locations are similar to that at the first location: if the dynamics is chaotic at the first site, the dynamics at the other sites are also chaotic. Hence, we do not plot the bifurcation diagrams for other locations. Our preliminary studies indicate that the range of the coupling parameter d, as a percentage of the interval [0,1], for which

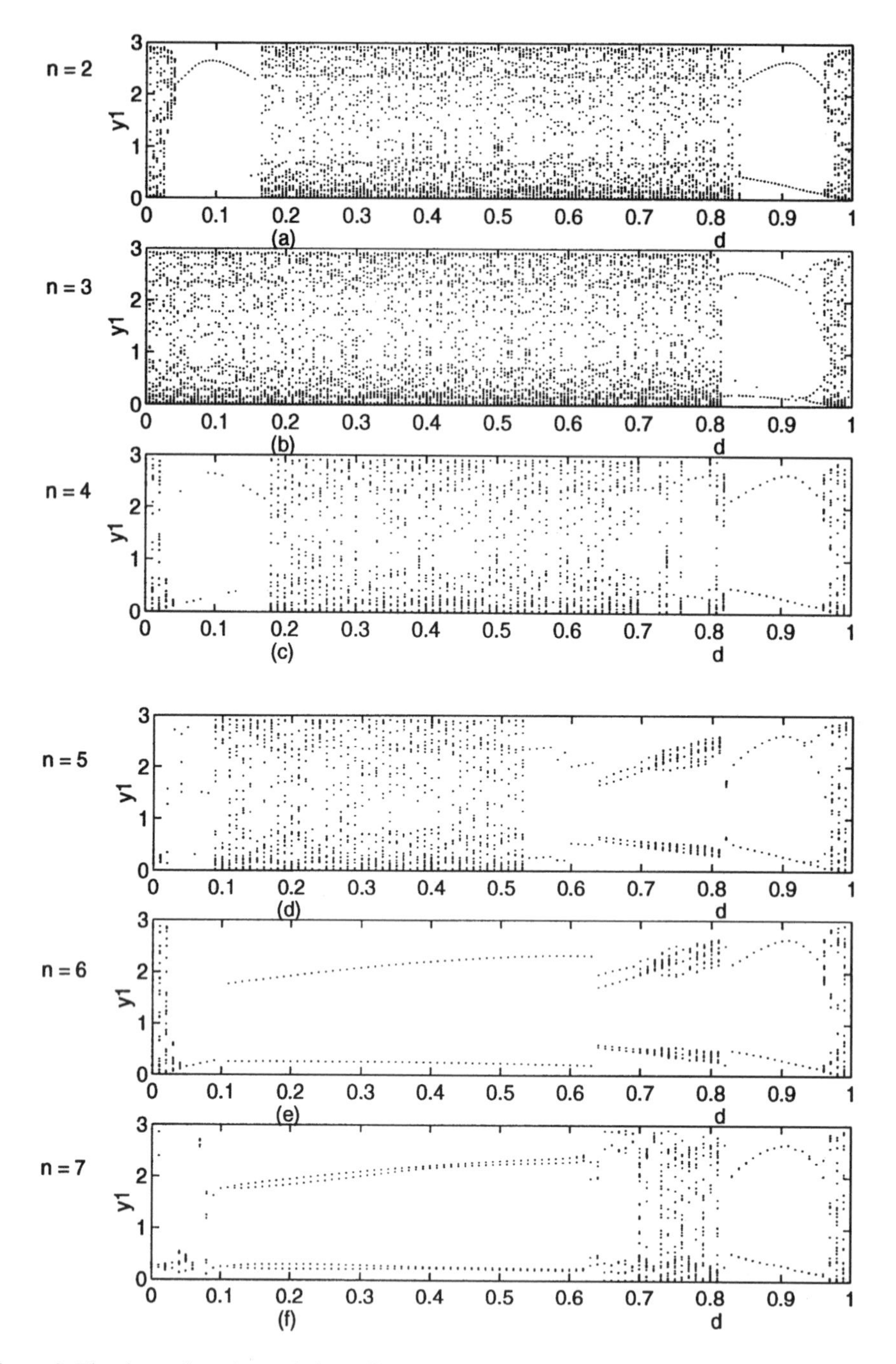

Figure 2 The dynamics of coupled nonlinear units as a function of the coupling parameter d. (a) number of units = 2. (b) number of units = 3. (c) number of units = 4 (d) number of units = 5. (e) number of units = 6. (f) number of units = 7. By and large, as the number of units increases the range of d for which stable behavior is attained increases.

stable dynamical behavior is attained tends to increase as the number of coupled systems increases. As n is increased to three the range of values of d (as a percentage of the interval [0,1]) for which stability is engendered actually decreases. But it increases rapidly as we continue to increase n and by $n = 6$ stable behavior is observed for most values of the coupling parameter d. We repeated this process for other values of the growth parameter r and found similar results; i.e., stability increased with n.

We also studied the effect of varying the values of the initial populations. However, as the size n increases the number of initial populations to be specified increases and as a result, we need to span a larger initial-condition-space. Consequently, it is difficult to span the entire initial-condition-space for $n > 3$. Therefore, these results should be treated as preliminary for $n > 3$. For the different initial conditions that we used, the trends in the stability engendered were similar: the percentage range of d for which stable dynamical behavior was observed, in general, increased as we increased the number of units that were coupled. Thus it can be conjectured that the overall stability of this nonlinear system increases as the size increases.

4 Discussion of Results

It is striking to note that coupling two chaotic systems can produce stable behavior. Thus migration tends to impose some sort of stability on highly nonlinear systems that are otherwise chaotic. Therefore coupling the adjacent habitats of a particular species may enhance a stable pattern in their population dynamics. Such enhancements may increase the survivability of rare animals as well. Further, the stability engendered increases as the size of the system increases.

More generally, we have demonstrated that *coupling two or more chaotic units can indeed stabilize all the units*. Cast differently, a "jug" of chaos when mixed with another "jug" of chaos can stabilize both. This stabilizing phenomenon may very well explain why there is so much stability in the physical world despite all the reported chaos. It may therefore be appropriate to *posit a hypothesis that highly nonlinear systems which are chaotic can be stabilized by coupling and it may be because of the coupling of nonlinear systems in real life that the physical world appears orderly*.

The stabilization demonstrated here is applicable to discrete dynamics in the form of maps. It is of interest to investigate whether the same stabilization trends hold for continuous systems as well. If so, it would further strengthen the conjectured hypothesis that coupling can stabilize otherwise chaotic nonlinear systems.

For the case of coupled tent maps Keller *et al.* [5] have shown that stable 2-period trajectories can be obtained by a suitable coupling of two tent maps that are chaotic but sufficiently close to 2-period motion. We, on the other hand, have demonstrated that periodic trajectories can be obtained by coupling chaotic exponential maps that are far away from periodic behavior.

References

[1] Chowdhury, R.A. and Chowdhury, K. (1991) Bifurcations in a coupled logistic map. *International Journal of Theoretical Physics,* **30**, 97.

[2] Fujisaka, H. and Yamada, T. (1982) Stability theory of synchronized motion. *Progress of Theoretical Physics*, **69**, 32.

[3] Gyllenberg, M., Soderbacka, G. and Ericsson, S. (1992) Does migration stabilize local population dynamics? Analysis of a discrete metapopulation model. *Mathematical Biosciences*, **118**, 25.

[4] Kaneko, K. (1983) Transition from torus to chaos accompanied by frequency lockings with symmetry breaking. *Progress in Theoretical Physics*, **69**, 1427.

[5] Keller, G. Kunzle, M. and Nowicki, T. (1992) Some phase transitions in coupled map lattices. *Physica D*, **59**, 39.

[6] May, R.M. (1976) Simple mathematical models with very complicated dynamics. *Nature*, **261**, 459.

[7] Pikovsky, P.A.S. and Grassberger. Symmetry breaking bifurcation for coupled attractors. *Journal of Physics*, A, **24**, 4587.

[8] Ricker, W.E. (1954) Stock and recruitment. *J. Fish. Rec. Board. Canada*, **11**, 559.

[9] Schuster, H.G., Martin, S. and Martienssen, W. (1986) New method for determining the largest Lyapunov exponent of simple nonlinear systems. *Physical Review A*, **33**, 3347.

[10] Udwadia, F.E. and Raju, N. (1997) Dynamics of coupled nonlinear maps and its application to ecological modeling. *Applied Mathematics and Computation*, **82**, 131.

[11] University of Southern California, Los Angeles. *Homer electronic database.*

18 ELECTING THE DIRECTORIAL COUNCIL

LEON A. PETROSJAN

Faculty of Applied Mathematics, St. Petersburg State University, Petrodvorets, 198904 St. Petersburg, Russia

1 Introduction

Suppose we have n enterprises, $A_1, \ldots, A_i, \ldots, A_n$, united in a concern. The employees of these enterprises elect the directorial council of the concern.

Denote by a_i the number of voters in A_i, $i \in \{1, \ldots, n\} = N$; a_i represents the power of A_i. Each enterprise can have at most one place in the council. For each i, a single candidate council member $b_i \in A_i$ is nominated. Thus, n candidates, $(b_1, \ldots, b_i, \ldots, b_n)$, are registered. Every voter votes (says "yes" or "no") for each of the candidates. For each voter his voting result is an n-dimensional vector whose ith component takes one of the two values, either "yes" (Y) or "no" (N), for example, $\underbrace{(Y, N, Y, \ldots, Y)}_{n}$. Candidate b_i is elected to the directorial council if he receives more than a half of positive votes, i.e., more than $\frac{1}{2} \cdot \sum_{i=1}^{n} a_i$ voters say "yes" for b_i. Suppose the directorial council (DC) $B = \{b_i : i \in S \subset N\}$ is elected. The DC B gets a payoff $K > 0$ which is independent of the number of members of DC and its staff, if $\sum_{i \in S} a_i > \frac{\sum_{i \in N} a_i}{2} = \frac{a(N)}{2}$. If $\sum_{i \in S} a_i \leq \frac{a(N)}{2}$ the payoff K is zero.

The amount K is distributed between the DC members proportionally to the power of their enterprises. Thus, the payoff to $b_i \in B$ (or what is the same, to b_i's company A_i) is given by

$$\beta_i = \frac{a_i}{\sum_{i \in S} a_i} \cdot K \tag{1}$$

Denote $\sum_{i \in S} a_i = a(S)$. Then (1) can be written as

$$\beta_i = \frac{a_i}{a(S)} \cdot K \tag{2}$$

For $i \notin S$ we have $\beta_i = 0$.

We pose the following questions: how the voters should vote, what is the optimal size of DC and what is the optimal DC membership. To find answers to these questions, we shall construct a game theoretic model and propose an approach leading to a Nash equilibrium in a specially constructed multistage game with complete information.

2 Simultaneous *n*-person Voting Game

Consider sets (coalitions) $\hat{S} \subset N$ satisfying

$$a(\hat{S}) > \frac{a(N)}{2} \tag{3}$$

where $a(\hat{S}) = \sum_{i \in \hat{S}} a_i$, $a(N) = \sum_{i \in N} a_i$. Every such $\hat{S}$ will be called *admissible*. Define $\bar{S}$, a *minimal admissible set*, by

$$a(\bar{S}) = \min_{\hat{S}} a(\hat{S}). \tag{4}$$

Here $\hat{S}$ runs through all admissible sets, i.e., subsets of N satisfying (3). The sets $\bar{B} = \{b_i, i \in \bar{S}\}$ and $\hat{B} = \{b_i, i \in \hat{S}\}$ will be called the *optimal DC* and *admissible DC*, respectively.

The coalition $\bar{S}$ comprises more than a half of the voters and is minimal among all coalitions with this property. It is clear that $\bar{S}$ defined by (3), (4) may not be unique. The enterprises in coalition $\bar{S}$ are not interested in letting any other enterprise join them because any extension of $\bar{S}$ would imply decrease in payoff to every member of $\bar{S}$ (see (1), (2)).

How should the members of $\bar{S}$ behave in order to guarantee that DC is formed of the candidates of $\bar{S}$?

Suppose that every company A_i tells its voters how to vote (the members of A_i form a voting coalition). If A_i for each $i \in \bar{S}$ prescribes "yes" for all candidates from $\bar{B}$, and "no" for all other candidates, the DC of $\bar{S}$ candidates is elected. Consider a simultaneous *n*-person voting game Γ with voters treated as players. The number of players in Γ is therefore $\bar{n} = a(N)$. Every player l has 2^n strategies. The strategy set of voter l consists of all vectors of the form $\alpha^l = \{\alpha_1^l, \ldots, \alpha_i^l, \ldots, \alpha_n^l\}$, where α_i^l is either "yes", or "no". In situation $\alpha = (\alpha^1, \ldots, \alpha^l, \ldots, \alpha^{\bar{n}})$ the result of voting is identified as follows: if the sum of "yes" at ith place in all players' strategies is more than $\frac{a(N)}{2}$, then the candidate of company A_i is elected to the DC, otherwise he is not elected. Suppose in situation α the DC $B = \{b_i : i \in S\}$ where S is an admissible coalition is elected. Then each voter l from A_i where $i \in S$ receives the amount

$$k_l(\alpha) = \frac{K}{\sum_{i \in S} a_i} = \frac{K}{a(S)}, \quad l = 1, \ldots, a(N), \tag{5}$$

the payoffs to all other voters' are zero. If S is not admissible, then $k_l(\alpha) = 0$ for all l.

Now we construct a Nash equilibrium in game Γ (J. Nash [1950]). Let $\bar{S}$ be defined by (3) and (4). Define strategy $\bar{\alpha}^l$ of player (voter) l as follows. If $l \in A_i$ for some $i \in \bar{S}$, then

$$\bar{\alpha}_i^l = \text{“yes”, if } i \in \bar{S},$$

$$\bar{\alpha}_i^l = \text{“no”, if } i \notin \bar{S}.$$

For $l \notin A_i$ where $i \in \bar{S}$ the strategy $\bar{\alpha}^l$ is arbitrary.

Theorem *The n-tuple of strategies $\bar{\alpha} = (\bar{\alpha}^1, \ldots, \bar{\alpha}^n)$ is a Nash equilibrium in Γ.*

Proof The payoffs $k_l(\alpha)$ are given by

$$k_l(\bar{\alpha}) = \frac{K}{a(\bar{S})}, \text{ for } l \in A_i, \; i \in \bar{S}, \tag{6}$$

$$k_l(\bar{\alpha}) = 0, \text{ for } l \in A_i, \; i \notin \bar{S}. \tag{7}$$

Take α_l, an arbitrary strategy of player l. Let us show that

$$k_l(\bar{\alpha} \| \alpha^l) \leq k_l(\bar{\alpha}), \; l = 1, \ldots, \; \bar{n} = a(N). \tag{8}$$

Here and in what follows, $\bar{\alpha}\|\alpha^l$ stands for $\bar{\alpha}$ with $\bar{\alpha}_l$ replaced by α_l. Suppose $l \in A_k$, $k \in \bar{S}$. If the change of strategy $\bar{\alpha}^l$ to α^l changes the result of the vote, there are two possibilities:

a) The candidate $b_k \in A_k$ is elected, i.e., $b_k \in \hat{B}$, where $\hat{B}$ is the DC elected in situation $(\bar{\alpha}\|\alpha^l)$. If $a(\hat{S}) > \frac{a(N)}{2}$ where $\hat{S} = \{i : b_i \in \hat{B}\}$, then

$$k_l(\bar{\alpha}\|\alpha^l) = \frac{K}{a(\hat{S})}$$

if $\hat{S}$ is admissible. Since $\hat{S}$ is not necessarily a minimal admissible set, $a(\hat{S}) \geq a(\bar{S})$, and (8) follows from (6). If $\hat{S}$ is not admissible, then by the definition $k_l = 0$.

b) The candidate $b_k \in A_k$ is not elected. Then $b_k \notin \hat{B}$, where $\hat{B}$ is the DC elected in situation $(\bar{\alpha}\|\alpha^l)$, and by definition we have $k_l(\bar{\alpha}\|\alpha^l) = 0$. Inequality (8) holds true.

Suppose now that $l \in A_k$, $k \notin \bar{S}$. In this case the change of the strategy by player l does not change the minimal admissible coalition and the DC. Thus, in situation $(\bar{\alpha}\|\alpha^l)$ company A_k is not represented in the DC, and we have $k_l(\bar{\alpha}\|\alpha^l) = 0$. The theorem is proved.

From the theorem it follows that for different minimal admissible coalitions $\bar{S}$ we get different Nash equilibria in Γ.

3 Multistage Game Generating a Minimal Admissible Coalition

Consider an n-person multistage game G with complete information [H. Kuhn, 1953] where enterprises $A_1, \dots, A_l, \dots, A_n$ act as players. The model will allow the minimal admissible coalition $\bar{S}$ to be formed dynamically. Player A_i will also be identified as i. Thus, in the notations of the previous section, $N = \{1, \dots, n\}$ is the set of all players.

It is easier to illustrate the game G by the flow chart shown in Fig. 1. The rectangles and rhomboids in the flow chart are explained as follows.

Rectangle 2. Here and in what follows M is the set of all active players who have not made their decisions yet, and S is the set of players who have agreed to join in a coalition. At the beginning, $M = N$, and S is empty. We write r for a stage number in the game. The game begins with $r = 1$.

Rectangle 3. "Random draw" means that all the players in M have equal probabilities of being the next decision maker.

Rectangle 4. Decision maker i is excluded from the set of active players.

Rhomboid 5. The decision maker decides whether to join coalition S, or not.

Rectangle 6. Player i joins coalition S.

Rhomboid 7. Decision maker i may be either the last active player, or not.

Rhomboid 8. Player i is the last active player. A check is made to see whether S is an admissible coalition, i.e. whether

$$a(S) > \frac{a(N)}{2}$$

Rectangle 9. Coalition S is admissible. Each player j receives his payoff h_j:

$$h_j = \frac{a_j}{a(S)} \cdot K \text{ if } A_j \in S;$$

$$h_j = 0 \text{ if } j \notin S.$$

Rhomboid 10. Player i is not the last active one. Then he either selects the next decision maker, i.e., invites another active player to join coalition S, or refuses to select the next decision maker.

Rectangle 11. Player i selects the next decision maker $j \in M$. Now j acts. A new stage begins.

Rhomboid 12. Player i refuses to select the next decision maker. The formation of coalition S is finished. A check is made to see whether S is an admissible coalition, i.e., whether

$$a(S) > \frac{a(N)}{2}.$$

Rhomboid 13. Decision maker i has not joined coalition S. A check is made to see whether S is admissible.

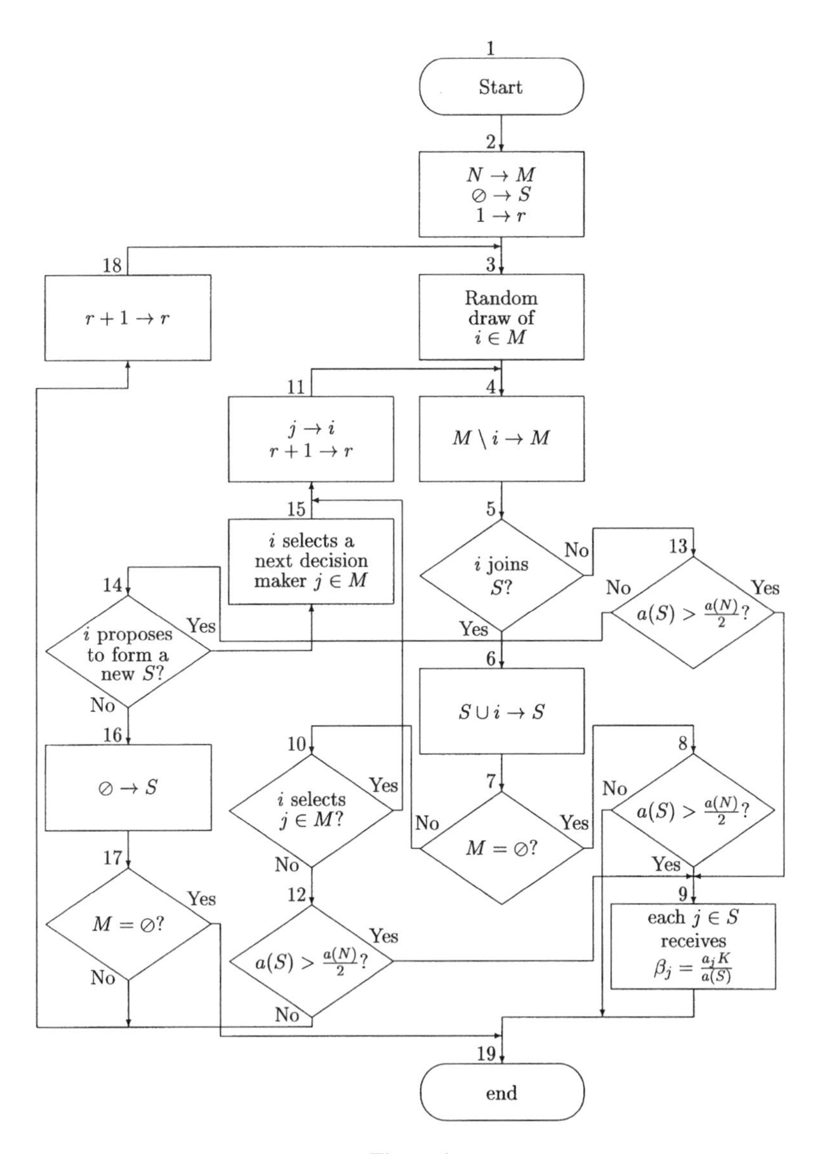

Figure 1

Rhomboid 14. The admissibility condition is not satisfied, i.e.,

$$a(S) \leq \frac{a(N)}{2}.$$

Now player i has two options: either to propose to create a new coalition including i and some of the remaining active players, or to abandon the game.

Rectangle 15. Player i decides to create a new coalition including i and some of the remaining active players, and selects the next decision maker $j \in M$.

Rectangle 16. Player i abandons the game. Coalition S disappears, $S = \oslash$.

Rhomboid 17. If i is the last active player, the game terminates. If not, the game continues with the current set of active players, M.

Rectangle 18. The next stage begins.

Positions. The initial position is $u_0 = N$, the set of all players in G. If player i is selected by a random draw or invited to join coalition S by a preceding decision maker, then the game position is the triple

$$u = (M, S, i),$$

where M is the set of all active players, S is the coalition being formed, and i is the decision maker.

If the preceding decision maker abandoned the game or refused to continue the formation of the coalition, then the game position is the triple

$$u = (M, S, R_i),$$

where M is the set of all active players, S is the coalition being formed, and R_i designates the negative decision of the preceding decision maker i.

Choice set $A(u)$. At the initial position u_0 the choice set $A(u_0)$ is the set of players, N. Each player is chosen with probability $\frac{1}{|N|}$. In position $u = (M, S, i)$ the choice set of the decision maker i comprises the following choices:

a) $\{R\}$, abandon the game;
b) $\{RY_k, k \in M\}$, refuse to join coalition S, decide to form a new coalition including i, and suggest some player $k \in M$ to join the latter coalition and be the next decision maker;
c) $\{YR\}$, accept the offer to join coalition S and refuse to invite any active player to join S;
d) $\{YY_k, k \in M\}$, agree to join coalition S and invite some player $k \in M$ to join S and be the next decision maker.

In position $u = (M, S, R_i)$ we have $A(u) = M$. If $S = \oslash$ or $|S| \leq \frac{a(N)}{2}$, the next decision maker comes from M randomly, according to the uniform probability distribution. If $|S| > \frac{a(N)}{2}$, $u = (M, S, R_i)$ is a terminal position ($A(u) = \oslash$).

The set of all positions that can follow choice $a \in A(u)$ at position u will be denoted by $D(u, a)$. The entry "end" in Fig. 1 indicates that $D(u, a)$ is empty. We write $D(u)$ for the union of all $D(u, a)$ as a runs through $A(u)$.

Histories A history q in G is a sequence of positions

$$q = (u_0, \ldots, u_T)$$

where

$$u_{t+1} \in D(u_t) \text{ for } t = 1, \ldots, T - 1.$$

The set of all histories q is denoted by Q.

Terminal choices. A choice $a \in A(u)$ is called terminal, if $D(u, a)$ is empty. Not every position u admits a terminal choice. A position u such that $A(u)$ contains at least one terminal choice is called terminable. The set of all terminable positions is denoted by u_T. A history $q = (u_0, \ldots, u_T)$ is called terminable, if its last position is terminable. The set of all terminable histories is denoted by Q_0.

Plays. A play z is defined to be a terminable history $q = (u_0, \ldots, u_T)$ together with a terminal choice $a_T \in A(u_T)$,

$$z = (u_0, \ldots, u_T; a_T).$$

The set of all plays is denoted by Z.

Payoffs for plays. Let z be a play and S be the coalition in the terminal position of z. Let S be admissible. Then $h_i(z)$, the payoff to player i for the play z, is given by (2) if $i \in S$, and zero otherwise. If S is not admissible, then $h_i(z) \equiv 0$ for all $i = 1, \ldots, n$.

Strategy. A strategy of player i, α^i, is a function assigning a single choice from the choice set $A(u)$ in every position u where i is the decision maker.

Payoffs in G. In situation $\alpha = (\alpha^1, \ldots, \alpha^i, \ldots, \alpha^n)$ the payoff to player i in game G, $k_i(\alpha)$, is defined to be the mathematical expectation of the payoffs $h_i(z)$ in individual plays z realized whenever the players use strategies $\alpha^1, \ldots, \alpha^i, \ldots, \alpha^n$.

We shall find a Nash (subgame perfect) equilibrium in game G [Selten, 1975].

Theorem *The following n-tuple of strategies $(\bar{\alpha}^1, \ldots, \bar{\alpha}^i, \ldots, \bar{\alpha}^n)$ is an absolute (subgame perfect) Nash equilibrium in G:*

Let $\bar{S}$ be a minimal admissible set. If $i \notin \bar{S}$, $\bar{\alpha}^i$ is arbitrary. Suppose $i \in \bar{S}$. If position $u = (M, S, i)$ (where $i \in \bar{S}$ is the decision maker) is such that $S \cap \bar{S} = \oslash$, then player i refuses to join coalition S, decides to create a new coalition including himself, invites another player from $\bar{S}$ to join the newly organized coalition, and suggests him to be the next decision maker. If $S \cap \bar{S} \neq \oslash$ and $(M \setminus i) \cap \bar{S} \neq \oslash$, player i agrees to join coalition S, invites some player $k \in (M \setminus \{i\}) \cap \bar{S}$ to join S, and suggests k to be the next decision maker. If $S \cap \bar{S} \neq \oslash$ and $(M \setminus \{i\}) \cap \bar{S} = \oslash$, player i agrees to join coalition S and does not invite any player from M to join S.

Proof Let us compute the payoffs in the situation $\bar{\alpha} = (\bar{\alpha}^1, \ldots, \bar{\alpha}^n)$. From the construction of strategies $\bar{\alpha}^i$, $i = 1, \ldots, n$ it follows that for each play z realized under $\bar{\alpha}$ the coalition $\bar{S}$ is formed in the terminal position of z. Hence the payoffs to the players are

$$k_i(\bar{\alpha}) = \frac{K \cdot a_i}{a(\bar{S})}, \text{if } i \in \bar{S}, \tag{11}$$

$$k_i(\bar{\alpha}) = 0, \text{if } i \notin \bar{S}.$$

Consider the situation $(\bar{\alpha} \| \alpha^i)$ where α^i is some strategy of player i. If $i \notin \bar{S}$, player i cannot prevent the formation of coalition $\bar{S}$ in every play arising in situation $(\bar{\alpha} \| \alpha^i)$. Thus

$$k_i(\bar{\alpha}) = k_i(\bar{\alpha} \| \alpha^i) = 0.$$

Now let $i \in \bar{S}$. By the definition of $\bar{S}$ β_i (2) is maximized at $S = \bar{S}$. Then in any play z realized under $(\bar{\alpha}\|\alpha^i)$ the payoff to player i, $h_i(z)$, does not exceed $k_i(\bar{\alpha})$. Hence

$$k_i(\bar{\alpha}\|\alpha^i) \leq k_i(\bar{\alpha}) \text{for} i \in \bar{S}.$$

The subgame perfectness of the situation $\bar{\alpha}$ can be proved in the similar way in any subgame of the game Γ. The theorem is proved.

4 Conclusion

In Sections 2 and 3 we described two ways of forming the DC representing a minimal admissible coalition $\bar{S}$. The situation where a number of DC members is prescribed and numbers of candidates from A_i's are not fixed, can be modeled by simultaneous and multistage games very similar to Γ and G described in Sections 2 and 3, respectively. In this situation, the main point is again identifying a minimal admissible coalition $\bar{S}$ that advises its employees to vote for the planned DC.

References

[1] Kuhn, H.W. (1953) Extensive games and the problem of information. *Annals of Mathematics Studies*, **28**, 193–216.

[2] Nash, J.F. (1950) Equilibrium points in n-person games. *Proc. Nat. Acad. Sci. USA*, **36**, 48–49.

[3] Selten, R. (1975) Reexamination of the perfectness concept for equilibrium points in extensive games. *International Journal of Game Theory*, **4**, 25–55.

[4] Selten, R. (1991) A Demand Commitment Model of Coalition Bargaining. Discussion paper N B-191, University of Bonn.

19 AN APPROACH TO BUILDING DYNAMICS FOR REPEATED BIMATRIX 2 × 2 GAMES INVOLVING VARIOUS BEHAVIOR TYPES*

ANATOLII F. KLEIMENOV

Institute of Mathematics and Mechanics, Ural Branch of Russian Academy of Sciences, 620219 Ekaterinburg, Russia

A new approach to building dynamics in repeated 2 × 2 bimatrix games originating from theory of closed-loop differential games is presented, and a natural formalism for viewing typical players' behaviors such as normal, altruistic, aggressive and paradoxical is discussed.

1 Introduction

There are various approaches to constructing dynamics in repeated games (see, e.g., [1]–[6]). Here we outline an approach originating from theory of closed-loop differential games, [7], [8], and, more specifically, a nonzero-sum branch of this theory, [9], [10]. The following points are characteristic in our approach:

(i) local displacements towards the states nonimprovable for one player and another (in the sense of Pareto),
(ii) the selection of a local displacement through finding a Nash equilibrium in some "local" bimatrix game;
(iii) a classification of such typical behaviors as normal, altruistic, aggressive, and paradoxical.

* The work presented in this paper has been supported by the Russian Fund for Fundamental Research (97-01-00161).

2 Bimatrix 2×2 Game

We deal with a mixed strategy bimatrix game with payoff matrixes

$$A = \begin{pmatrix} a_{11} & a_{12} \\ a_{21} & a_{22} \end{pmatrix}, \quad B = \begin{pmatrix} b_{11} & b_{12} \\ b_{21} & b_{22} \end{pmatrix} \tag{1}$$

of players 1 and 2, respectively. Note that mixed strategies arise naturally when the game is assumed to be played out between agents from two large groups. If $(p, 1-p)$, $0 \leq p \leq 1$, is a mixed strategy of player 1 and $(q, 1-q)$, $0 \leq q \leq 1$, a mixed strategy of player 2, then p and q are interpreted as shares of agents in the first and, respectively, second group who play the pure strategy 1. Payoffs to players 1 and 2 are given, respectively, by

$$\begin{aligned} f_1(p,q) &= Cpq - c_1 p - c_2 q + a_{22}, \\ f_2(p,q) &= Dpq - d_1 p - d_2 q + b_{22}, \end{aligned} \tag{2}$$

where

$$\begin{aligned} C &= a_{11} - a_{12} - a_{21} + a_{22}, \quad c_1 = a_{22} - a_{12}, \quad c_2 = a_{22} - a_{21}, \\ D &= b_{11} - b_{12} - b_{21} + b_{22}, \quad d_1 = b_{22} - b_{12}, \quad d_2 = b_{22} - b_{21}. \end{aligned} \tag{3}$$

Pair (p,q) in the unit square

$$E = \{(p,q): \quad 0 \leq p \leq 1; \quad 0 \leq q \leq 1\} \tag{4}$$

characterizes a *state* of the players (or the associated groups of agents).

3 Repeated Bimatrix 2×2 Game

Let the bimatrix game be repeated N times, where N is large. Let $(p_0, q_0) \in E$ be an initial state. To define a dynamics of states, or, equivalently, mixed strategies in the repeated bimatrix game, we introduce players' control and information areas and specify their objectives. Assume that in round $k+1$ ($k = 0, 1, ..., N-1$) the players are allowed to pass from the current state $(p_k, q_k) \in E$ to any state $(p_{k+1}, q_{k+1}) \in M_{\alpha_1,\alpha_2}(p_k, q_k)$ where

$$M_{\alpha_1,\alpha_2}(p_k, q_k) = \{(p,q) \in E : |q - q_k| \leq \alpha_1, \ |q - q_k| \leq \alpha_2\}. \tag{5}$$

Here α_1 and α_2 are positive and sufficiently small. Small α_1 and α_2 indicate that the inner structure of the groups of interacting agents evolves slowly enough. Thus, $M_{\alpha_1,\alpha_2}(p_k, q_k)$ characterizes control abilities of players 1 and 2 at state (p_k, q_k). We call the projections of $M_{\alpha_1,\alpha_2}(p_k, q_k)$ to axis p and q *control areas* of players 1 and 2, respectively (at state (p_k, q_k)). A game dynamics as a control process is described by

$$\begin{aligned} p_{k+1} &= p_k + u_k, \quad u_k \in [p_k - \alpha_1, p_k + \alpha_1] \cap [0,1], \\ q_{k+1} &= q_k + v_k, \quad v_k \in [q_k - \alpha_2, q_k + \alpha_2] \cap [0,1], \end{aligned} \tag{6}$$

or, in continuous time, by

$$\begin{aligned} \dot{p} &= \gamma u, \quad u \in [p - \alpha_1, p + \alpha_2] \cap [0, 1], \\ \dot{q} &= \gamma v, \quad v \in [q - \alpha_2, q + \alpha_2] \cap [0, 1], \end{aligned} \tag{6}$$

where $\gamma > 0$. Note that approximate discrete time Euler spline motions (see [8]) for system (7) with time step $\Delta t = 1/\gamma$ are obviously governed by (6). We deal with the discrete time dynamics (6) only.

Let us characterize information available to the players. At a current state (p_k, q_k) each player knows, first, (p_k, q_k) and, second, the values of the payoff functions f_1 and f_2 (2) in a rectangle $M_{\beta_1,\beta_2}(p_k, q_k)$ where $\beta_1 \geq \alpha_1$ and $\beta_2 \geq \alpha_2$. We call $M_{\beta_1,\beta_2}(p_k, q_k)$ *the information area* (at (p_k, q_k)). Two opposite extremes correspond to $(\beta_1, \beta_2) = (\alpha_1, \alpha_2)$, and $M_{\beta_1,\beta_2} = E$. These are the cases of *local* and *global information*. All other intermediate situations can be qualified as *incomplete information*.

Now we specify the objectives of the players. We assume two types of objectives, long-term and short-term. A long-term objective of player 1 (respectively, player 2) is to maximize $f_1(p_N, q_N)$ (respectively, $f_2(p_N, q_N)$), a payoff in the final round N. A short-term goal of player 1 (respectively, player 2) in round k is to maximize his (or her) current payoff, $f_1(p_k, q_k)$ (respectively, $f_2(p_k, q_k)$). Note that in the case of local information each player can actually pursue only his (or her) short-term objective. In a general situation, short- or long-term objectives, or a mixture of both motivate players' decisions.

To specify behavior patterns incorporating players' objectives, we invoke some ideas from theory of nonzero-sum closed-loop differential games. These are, first, the idea of "nondeterioration" of guaranteed results (in local or global understanding) for both players (see [9], [10]), second, the idea of maximum displacement (see [5], [6]) toward a Pareto optimal state, and, third, the idea of finding local transitions via a search of Nash equilibria in appropriate "local" games.

Another element of the approach discussed here is a formalism for the analysis of such typical behaviors of the players as normal, aggressive, altruistic, and paradoxical.

4 Noncooperative Dynamics

In what follows, for the sake of brevity we focus on the case of local information. So, we assume that

$$M_{\alpha_1,\alpha_2}(p, q) = M_{\beta_1,\beta_2}(p, q)$$

for all states (p, q).

Let (p_k, q_k) be a current state in the repeated game. We assume that the players find a transition to a next state, (p_{k+1}, q_{k+1}), via the analysis of their "local" extremal problems:

Problem 1 *Find* p^1 *such that*

$$(p^1, q_k) = \underset{(p,q_k)\in M_{\beta_1,\beta_2}(p_k,q_k)}{\arg\max} f_1(p, q_k) \tag{8}$$

under the condition

$$f_2(p, q_k) \geq f_2(p_k, q_k). \tag{9}$$

Problem 2 Find q^2 such that

$$(p_k, q^2) = \underset{(pk,q)\in M_{\beta_1,\beta_2}(p_k,q_k)}{\arg\max} f_2(pk, q) \tag{10}$$

under the condition

$$f_1(p_k, q) \geq f_1(p_k, q_k). \tag{11}$$

In Problem 1 player 1 finds an optimal point p^1 with respect to his (or her) payoff over the information area $M_{\beta_1,\beta_2}(p_k, q_k)$ under the condition that player 2 does not lose in payoff provided he (or she) does not change his (or her) current state component q_k, or, equivalently, his (or her) current mixed strategy. Problem 2 is interpreted analogously. Note that players solve their local Problems 1 and 2 separately; in other words, they do not cooperate in finding their new candidate strategies, p^1 and q^1. To identify (p_{k+1}, q_{k+1}), the players proceed as follows. They check whether one of the equalities

$$f_1(p^1, q_k) = f_1(p_k, q_k), \tag{12}$$

$$f_2(p_k, q^2) = f_2(p_k, q_k) \tag{13}$$

holds. If both (12) and (13) hold, then the players stay at (p_k, q_k), i.e., set $(p_{k+1}, q_{k+1}) = (p_k, q_k)$. If (12) holds and (13) does not hold, then $(p_{k+1}, q_{k+1}) = (p_k, q^2)$. Symmetrically, if (13) holds and (12) does not hold, then $(p_{k+1}, q_{k+1}) = (p^1, q_k)$. Finally, if neither (12) nor (13) hold, then the players play a local pure strategy bimatrix game in which the payoff matrixes of players 1 and 2 are given, respectively, by

$$A^{nc} = \begin{pmatrix} f_1(p_k, q_k) & f_1(p_k, q^2) \\ f_1(p^1, q_k) & f_1(p^1, q^2) \end{pmatrix}, \quad B^{nc} = \begin{pmatrix} f_2(p_k, q_k) & f_2(p_k, q^2) \\ f_2(p^1, q_k) & f_2(p^1, q^2) \end{pmatrix}. \tag{14}$$

In this local bimatrix game, strategy 1 is repeating p_k for player 1 and repeating q_k for player 2, and strategy 2 is switching from p_k to p^1 for player 1 and switching from q_k to q^2 for player 2. To identify (p_{k+1}, q_{k+1}), the players find a pure strategy Nash equilibrium, *NE*, in the local bimatrix game. One can prove that this game always has a pure strategy *NE*; moreover, four situations for *NE*, and, respectively, (p_{k+1}, q_{k+1}), may occur. Namely, the next cases are the only admissible ones:

1) $(1, 2)$ is a single *NE*, and $(p_{k+1}, q_{k+1}) = (p_k, q^2)$,
2) $(2, 1)$ is a single *NE*, and $(p_{k+1}, q_{k+1}) = (p^1, q_k)$,
3) $(2, 2)$ is a single *NE*, and $(p_{k+1}, q_{k+1}) = (p^1, q^2)$,
4) $(1, 2)$ and $(2, 1)$ are two *NE*, and (p_{k+1}, q_{k+1}) is either (p_k, q^2), or (p^1, q_k), with some (possibly equal) probabilities.

The described game dynamics can be entirely investigated analytically. A detailed analysis (as well as proofs omitted here) will be provided in further publications.

In the case of incomplete information a point characterized above as (p_{k+1}, q_{k+1}) has a sense of a project location for the next state; we denote it (p^*_{k+1}, q^*_{k+1}). An actual (p_{k+1}, q_{k+1}) is defined as a point closest to (p^*_{k+1}, q^*_{k+1}) in the intersection $M_{\alpha_1,\alpha_2}(p_k, q_k) \cap I_k$ where I_k is a segment connecting (p_k, q_k) to (p^*_{k+1}, q^*_{k+1}).

5 Cooperative Dynamics

Replacing Problems 1 and 2 by the next Problems 3 and 4, we come to a cooperative decision-making pattern in round-to-round transitions.

Problem 3 *Find* (p^1, q^1) *such that*

$$(p^1, q^1) = \underset{(p,q)\in M_{\beta_1,\beta_2}(p_k,q_k)}{\arg\max} f_1(p, q) \tag{15}$$

under the condition

$$f_2(p, q) \geq f_2(p_k, q_k). \tag{16}$$

Problem 4 *Find* (p^2, q^2) *such that*

$$(p^2, q^2) = \underset{(p,q)\in M_{\beta_1,\beta_2}(p_k,q_k)}{\arg\max} f_2(p, q) \tag{17}$$

under the condition

$$f_2(p, q) \geq f_2(p_k, q_k). \tag{18}$$

We see that the players solve Problems 3 and 4 jointly.

A decision-making pattern is as follows. The players check if one of the equalities

$$f_1(p^1, q^1) = f_1(p^2, q^2), \tag{19}$$

$$f_2(p^1, q^1) = f_2(p^2, q^2) \tag{20}$$

holds. If both (19) and (20) hold, then $(p_{k+1}, q_{k+1}) = (p^1, q^1)$ or $(p_{k+1}, q_{k+1}) = (p^2, q^2)$. If (19) holds and (20) does not hold, then $(p_{k+1}, q_{k+1}) = (p^2, q^2)$. Symmetrically, if (20) holds and (19) does not hold, then $(p_{k+1}, q_{k+1}) = (p^1, q^1)$. Finally, if neither (19) nor (20) hold, then the players play a pure strategy local bimatrix game in which the payoff matrixes of players 1 and 2 are, respectively,

$$A^c = \begin{pmatrix} f_1(p^1, q^1) & f_1(p^1, q^2) \\ f_1(p^2, q^1) & f_1(p^2, q^2) \end{pmatrix}, \quad B^c = \begin{pmatrix} f_2(p^1, q^1) & f_2(p^1, q^2) \\ f_2(p^2, q^1) & f_2(p^2, q^2) \end{pmatrix}. \tag{21}$$

In this local bimatrix, strategy i $(i = 1, 2)$ of player 1 is switching from p_k to p^i, and strategy i of player 2 is switching from q_k to q^i. A structure of this local bimatrix game is identical to that described in the previous section. Correspondingly, a next state, (p_{k+1}, q_{k+1}), can be characterized.

The analysis of the global game trajectories under the noncooperative and cooperative dynamics lies beyond the scope of this paper. Preliminary arguments and simulations show that the cooperative dynamics is much richer in trajectory types than the noncooperative one. The consideration of incomplete and global information add new dynamical phenomena. For example, a game trajectory can stay at some point in the case of local information and pass it by in the case of global or incomplete information. An explanation is that a narrow information area may make the players act as antagonists, whereas a wider information horizon reduces antagonism and allows the players to find a compromise.

6 Behavior Types

Until now we assumed that each player is *normal* (*nr*) in the sense that his (or her) behavior is aimed at maximizing his (or her) own payoff. However, there might be other behavior types such as *altruistic* (*al*) ("the better for my rival, the better for me"), *aggressive* (*ag*) ("the worse for my rival, the better for me"), and *paradoxical* (*pr*) ("the worse for me, the better for me"). These three behavior types can be formalized in the following way.

Let us say that player 1 demonstrates

(i) altruistic behavior whenever he (or she) identifies his (or her) payoff matrix with B,
(ii) aggressive behavior whenever he (or she) identifies his (or her) payoff matrix with $-B$, and
(iii) paradoxical behavior whenever he (or she) identifies his (or her) payoff matrix with $-A$.

For player 2 we assume symmetric definitions.

Table 1

	nr	al	ag	pr
nr	(A, B)	(A, A)	$(A, -A)$	$(A, -B)$
al	(B, B)	(B, A)	$(B, -A)$	$(B, -B)$
ag	$(-B, B)$	$(-B, A)$	$(-B, -A)$	$(-B, -B)$
pr	$(-A, B)$	$(-A, A)$	$(-A, -A)$	$(-A, -B)$

These definitions indicate extremes in players' behaviors. Real individuals behave partially normally, partially altruistically, partially aggressively, and partially paradoxically. In other words, mixtures of behavior types would better agree with real dynamics.

If we restrict each player to "pure" behavior types, we arrive at 16 admissible combinations as shown in Table 1. In 4 combinations players' interests coincide, and they solve problems of optimal choice. In another 4 combinations the players have opposite interests and, thus, play zero-sum matrix games. The remaining 8 pairs determine nonzero-sum bimatrix games.

In the previous sections noncooperative and cooperative algorithms for building *normal–normal* game dynamics were defined. Replacing the original matrix pair (A, B) in these definitions by the matrix pairs given in Table 1, we obviously introduce 15 new types of game dynamics in both noncooperative and cooperative variants. If, moreover, the players moving along a (noncooperative or cooperative) game trajectory are allowed to switch their behavior types over time, mixed-behavior game processes arise. In this manner we naturally come to such a qualification of behaviors as, say, 80 percent normal and 20 percent altruistic, etc. A detailed analysis of mixed-behavior game processes will be presented in further publications.

7 Examples

In this section several simulation results are presented. The author is grateful to Mrs. L.V. Kukushkina who created a (C language) program for the simulation of the noncooperative and cooperative game dynamics.

Example 1 Consider the repeated Prisoner's Dilemma game with payoff matrixes

$$A = \begin{pmatrix} -5 & 0 \\ -10 & -1 \end{pmatrix}, \quad B = \begin{pmatrix} -5 & -10 \\ 0 & -1 \end{pmatrix}.$$

In this game the origin in the state space, $(p, q) = (0, 0)$, corresponds to a pure strategy Pareto solution, known as *cooperation*. We denote it C. Point C providing each player with -1 is obviously most favorable for the players as a final state but as

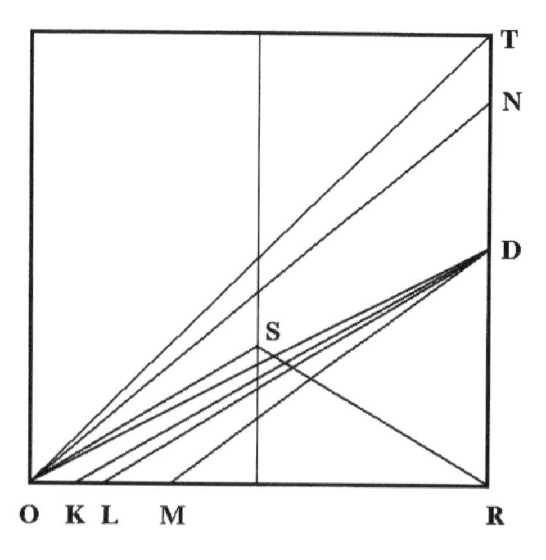

Figure 1

they start far away from C, it is problematic whether they can find an acceptable path to C.

It can be shown that all noncooperative trajectories (i.e., trajectories under the noncooperative dynamics) are stationary. Therefore, in what follows, we deal with the cooperative trajectories.

The simulated trajectories are shown in Fig. 1. We see that the trajectories starting at the diagonal OT or in its neighborhood reach C. Consider a trajectory starting at point D ($p_0 = 1,\ q_0 = 0.5$). In the case of local information with $\alpha_1 = \alpha_2 = \beta_1 = \beta_2 = 0.001$ the trajectory stops at point M. In the case of incomplete information with $\alpha_1 = \alpha_2 = \beta_1 = 0.001,\ \beta_2 = 0.016$ the trajectory stops at L, that is, closer to C than M. For $\alpha_1 = 0.001,\ \alpha_2 = 0.005,\ \beta_1 = 0.001,\ \beta_2 = 0.016$ the trajectory stops at K, that is, closer to C than L.

For $\alpha_1 = \alpha_2 = \beta_2 = 0.001, \beta_1 = 0.002$, a trajectory starting at point R ($p_0 = 1,\ q_0 = 0$) is stationary. However, if player 1 acts altruistically during some initial time period, the trajectory moves away from R and reaches point S. At S we let player 1 switch back to normal behavior, and set $\alpha_1 = \alpha_2 = \beta_1 = 0.001$, $\beta_2 = 0.019$. The trajectory continues and reaches cooperation, C.

Example 2 Consider a repeated bimatrix game with payoff matrixes

$$A = \begin{pmatrix} 3 & 2 \\ 1 & 4 \end{pmatrix},\ B = \begin{pmatrix} 3 & 4 \\ 2 & 1 \end{pmatrix}.$$

This game has a single mixed strategy Nash equilibrium, $p^N = 1/2,\ q^N = 1/2$. Let us examine situations in which closed, or cycling, trajectories can exist. If both players behave normally, neither the noncooperative nor cooperative dynamics can produce

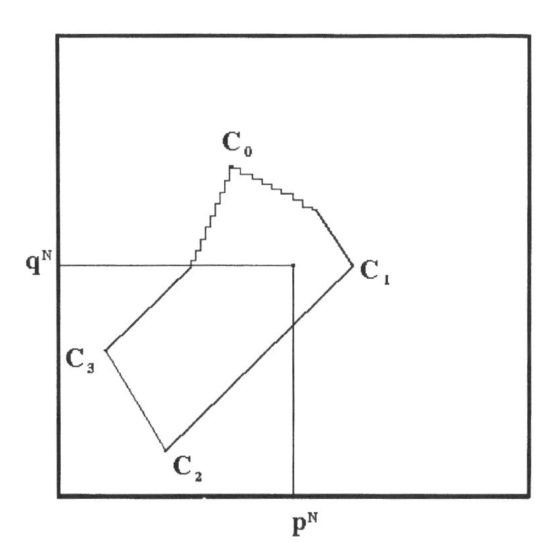

Figure 2

cycles. On the other hand, appropriate switchings in behavioral types allow cycles to be created under the cooperative dynamics. A corresponding trajectory is shown in Fig. 2. We set $\alpha_1 = \alpha_2 = \beta_1 = \beta_2 = 0.01$ (thus we have the case of local information) and start the trajectory at point C_0 $(p_0 = 0.38,\ q_0 = 0.68)$. Let the players behave normally until reaching point C_1 $(p_0 = 0.65,\ q_0 = 0.50)$. At this point player 2 switches to altruistic behavior. The trajectory reaches C_2 $(p_0 = 0.25,\ q_0 = 0.10)$ where both players switch to paradoxical behavior. The trajectory moves to C_3 $(p_0 = 0.14,\ q_0 = 0.33)$. At this point player 2 switches again to normal behavior, and the trajectory comes back to the initial point C_0.

8 Some Open Problems

Let us indicate three problems that (among others) can be associated with the suggested approach to forming dynamics in repeated 2 × 2 bimatrix games.

1. Given a desired trajectory in a repeated game, identify mixed behavioral types for the players allowing them to follow this trajectory.
2. Given a set of behavior mixes solving problem 1, find a mix with a maximum share of normal behavior.
3. Given a trajectory in a repeated game and a mix of players' behavioral types, identify payoff matrixes A and B.

References

[1] Maynard Smith, J. (1982) *Evolution and the Theory of Games*. Cambridge Univ. Press.
[2] Hofbauer, J. and Sigmund, K. (1988) *The Theory of Evolution and Dynamic Systems*. Cambridge Univ. Press, Cambridge.
[3] Friedman, D. (1991) Evolutionary games in economics. *Econometrica*, **59**(3), 637–666.
[4] Young, P. (1993) The evolution of conventions. *Econometrica*, **61**(1), 57–84.
[5] Kryazhimskii, A.V. and Osipov Yu.S. (1995) On differential-evolutionary Games. *Proceedings of the Steklov Institute of Mathematics*, **211**, 234–261.
[6] Tarasyev, A.M. (1994) A differential model for a 2×2 evolutionary game dynamics. International Institute for Applied Systems Analysis, Laxenburg, Austria, Working Paper, WP-94-63.
[7] Krasovskii, N.N. (1985) *Control of a Dynamical System*. Nauka, Moscow (in Russian).
[8] Krasovskii, N.N. and Subbotin, A.I. (1988) *Game-Theoretical Control Problems*. Springer, NY, Berlin.
[9] Kleimenov, A.F. (1993) *Nonantagonistic Positional Differential Games*. Nauka, Ekaterinburg (in Russian).
[10] Kleimenov, A.F. (1995) An approach to defining solution concept in N-person nonantagonistic positional differential games. *Int. Yearbook on Game Theory*, **3**.

INDEX